A FIELD GUIDE TO THE SMALLER BRITISH LEPIDOPTERA

ACKNOWLEDGMENT

The Society gratefully acknowledges
a grant of
£500
and an interest-free loan of £1000
from The Royal Society
towards the cost of this publication

A FIELD GUIDE TO THE SMALLER BRITISH LEPIDOPTERA

by

MEMBERS OF THE BRITISH ENTOMOLOGICAL & NATURAL HISTORY SOCIETY

Edited by

A.M. EMMET

Published by

The British Entomological & Natural History Society

at the Rooms of the Alpine Club,

74, South Audley Street, London, W1Y 5FF

ISBN No. 0 9502891 1 6

To

THE MEMORY OF L.T. FORD,

whose

A GUIDE TO THE SMALLER BRITISH LEPIDOPTERA

is the foundation

upon which this work has been built.

CONTENTS, AUTHORS AND THE PRINCIPAL PUBLISHED AND MANUSCRIPT WORKS BY OTHER WRITERS USED AS SOURCES.

INTRODUCTION

Since its publication in 1949, L. T. Ford's *A Guide to the Smaller British Lepidoptera* has been a standard work for the microlepidopterist and is still in steady demand. Stocks now being virtually exhausted, the Council of the British Entomological & Natural History Society considered the alternatives of reprinting the 1949 edition, or commissioning a new work which would follow modern taxonomic arrangement, incorporate species added to the British list since 1949 and include fresh information about those already described. The second alternative was chosen and this book is the result. Because of its many differences, it bears a new title; nevertheless, it is firmly based on Ford's foundation and his wording has been retained wherever it is suitable. Moreover, additional information is included which was collected by Ford in person and entered in his working copy of the "Guide", now in possession of Dr. J.D. Bradley.

Extensive research into the British Lepidoptera is currently being undertaken by a team of authors at work on *The Moths and Butterflies of Great Britain and Ireland,* and the services of these specialists has as far as possible been utilised. Readers will soon notice that work is more advanced in some families than others; it is not a reflection on an author if he is still in the early stages of his research and has little to add. The Society decided to go ahead at once with the "Field Guide", using the information so far available, however incomplete that might be.

Several books on the Microlepidoptera, published since 1949, are important sources. Foremost amongst these are *The Moths and Butterflies of Great Britain and Ireland,* Volume I, edited by J. Heath et al., *British Pyralid and Plume Moths,* by B.P. Beirne, *The Elachistidae (Lepidoptera) of Fennoscandia and Denmark,* by E. Traugott-Olsen and E. Schmidt Neilsen and *British Tortricoid Moths,* Volume I by J.D. Bradley, W.G. Tremewan and A. Smith, and Volume II by J.D. Bradley, W.G. Tremewan and B. Hargreaves; this last volume is still in typescript and I am privileged to have been allowed to make use of the rich fund of information it contains. Material has been extracted from various journals and authors have included their own field notes as well as those of their friends. Where records of life histories made in Britain are lacking or incomplete, material from continental authors has been included; such entries are indicated by an asterisk. In cases where a species has recently been added to the British list and is not described in a standard text-book, a reference is given to a relevant paper, wherever possible one in a British journal. The evidence for the occurrence of species which have dubious titles to their place on the British list is also stated.

This book is written for the field-worker. Its purpose is to enable him to find and recognise a species he is interested in from its field characteristics and to tell him what its requirements are if he wishes to rear it. The descriptions given, therefore, are not of the insect itself, but of its behaviour and the situation in which it may be found. An exception is the partial description of larvae given in certain instances where two or more species are to be found feeding in the same manner on the same foodplant. The reader might wish that this aid to identification had been more widely provided, but the necessary information was not always available. Moreover, in some families such as the Tortricidae the larvae are too variable for descriptions to be of much help. These partial descriptions are intended to be read in conjunction with those of the similar species. Other details, too, are sometimes included which will only be seen to be relevant when contrasted with those given for another species on the same plant (see, for example, nos 41 and 230). The reader must accept that it will not always be possible to ascribe the larva he finds to the exact species, though it is hoped that the field of choice will be narrowed.

The abbreviations 'O.', 'L.', 'P.' and 'I.', standing for 'Ovum', 'Larva', 'Pupa' and 'Imago', are here treated as singular. Though professedly these were also singular in the "Guide", Ford often unwittingly slipped into the plural with ambiguous or misleading results: for example, there is a difference between stating that a larva feeds "in shoots" and "in a shoot". My attempt to treat "Ovum" as singular was well-intended but, as I now see, misguided, and I apologise for clumsy phraseology under this heading. With some species I have reworded the entry so that "Ova" is the assumed subject. The naming of foodplants on the "O." line instead of the "L." line was done for economy of space; in the "Guide", the "O." line is generally wasted.

Nomenclature.

(a) *Lepidoptera.* The nomenclature used is basically that of Kloet & Hincks (1972). However, since its publication certain changes and additions have been made; these are being incorporated in a label list/log book by the same authors. The revised list is in press and should be available well before the "Field Guide" is published. We have here incorporated the amendments and wish to thank Dr. J.D. Bradley for letting us consult the new list while it was still in typescript. The "Field Guide" is up to date to the 1st of January, 1979. Intentional deviations are few. We have put *Coleophora badiipennella* (Duponchel) two places later so that it does not interrupt the sequence of the three species formerly regarded as one under the name *Coleophora limosipennella* (Duponchel). Several controversial "species" which are reduced to synonymy in the label list/log book are here listed separately; the reader is left to decide for himself on their status. They include (1082) *Ancylis diminutana* (Haworth); (1083) *A. subarcuana* (Douglas); (1150) *Epinotia cirsiana* (Zeller); (1167) *Eucosma fulvana* (Stephens); and (1205) *Pammene herrichiana* (Heinemann).

Where there has been a change of name from Kloet & Hincks, the superseded name is given as a synonym. The only other synonyms included are the names used by Ford, with the author's name amended, in some instances, to comply with Kloet & Hincks. Ford's serial number for each species is given to enable reference to be made, not only to the "Guide", but also to *A Revised Handbook of British Lepidoptera* by E. Meyrick, whose nomenclature and taxonomic arrangement Ford followed. Subfamilies are named only in the Gracillariidae and Tortricidae, where there is a change of authorship for the description of the species which follow.

(b) *Foodplants.* Ford used partly Latin and partly English names for foodplants; here Latin names are used throughout, but their English equivalents will be found in the Index to Foodplants, etc.. The names in both languages are those given in *English Names of Wild Flowers: A List Recommended by the Botanical Society of the British Isles* by J.G. Dony, F.H. Perring and C.M. Rob (1974). A few familiar but superseded names have been added as synonyms in the Index.

Where the generic name only of a foodplant is given, it indicates that the larva feeds on all (or the only) species in the genus. The generic name followed by "spp." denotes that the larva feeds on some, but not necessarily all, members of that genus. Important instances of this convention are "*Betula* spp,", here excluding *B. nana;* "*Quercus* spp.", here excluding *Q. ilex;* and "*Ulmus* spp.", here excluding *U. glabra.* I have found it particularly difficult to decide between "*Ulmus*" and "*Ulmus* spp.", since authors who do not use this convention give "*Ulmus*" as a foodplant without considering whether wych elm is acceptable. Wych elm is, in fact, distasteful to many species; leaf-mines found on that foodplant are generally aborted.

Acknowledgements.

Financial support has generously been given by S.N.A. Jacobs and R.S. Tubbs. The Council of the British Entomological & Natural History Society has also allocated a legacy from Sir Eric Ansorge to help defray the cost of publication.

The laborious task of typing has been undertaken by the following members of the Society:- D.J.L. Agassiz; E.S. Bradford; G.N. Burton; J.M. Chalmers-Hunt; R. Dyke; S.N.A. Jacobs; and P.A. Sokoloff.

I am also indebted to P.J. Johnson for his meticulous checking of the typescript, particularly of the names of species and foodplants, serial numbers and indices, which has greatly lightened my editorial burden.

ABBREVIATIONS, SYMBOLS, ETC..

1-12.	January-December.
O.	Ovum or Ova.
L.	Larva.
P.	Pupa.
I.	Imago.
*	An indication that the information which follows has been taken from continental authors.
	The serial number of each species is followed by Ford's serial number in parentheses.

Authors. The abbreviations of authors' names are those used in Meyrick's *A Revised Handbook of British Lepidoptera* (1928), with the following additions:-

D. & S.	Denis & Schiffermüller. Their names are placed in square brackets because their work was published anonymously.
H. & W.	Humphreys & Westwood.
L. & Z.	Lienig & Zeller.
P. & M.	Pierce & Metcalfe.
R. & T.	Richards & Thomson.
	Dates are added after an author's name only when he used the same name for two species in different works.

References. References are made to the following journals; the abbreviations which have been used are shown.

	Antenna.
Bull. amat. ent. Soc.	Bulletin of the Amateur Entomologists' Society.
Bull. ent. Res.	Bulletin of Entomological Research.
Dt. ent. Z.	Deutsche Entomologische Zeitschrift.
Ent.	Entomologist.
Ent. Gaz.	Entomologist's Gazette.
Ent. mon. Mag.	Entomologist's Monthly Magazine.
Ent. Rec.	Entomologist's Record and Journal of Variation.
Exot. Micro.	Exotic Microlepidoptera.
	Microscope.
Opusc. ent.	Opuscula Entomologica.
Proc.Brit.ent.nat.Hist.Soc.	Proceedings of the British Entomological & Natural History Society.
Proc.S.Lond.ent.nat.Hist.Soc.	Proceedings of the South London Entomological & Natural History Society.
	Svensk Insektfauna.
Trans.Soc.Brit.Ent.	Transactions of the Society for British Entomology.
Z.wein.ent.Ges.	Zeitschrift der Wiener Entomologischen Gesellschaft.

MICROPTERIGIDAE

Life history is almost unknown, and no differentiating characters can be given for the early stages. The ova are laid amongst vegetation on the surface of the soil. The larvae probably feed on leaf-litter, but possibly on fungal hyphae. The pupae are enclosed in tough, silken cocoons. The imagines feed on the pollen of various herbaceous plants, shrubs and trees, showing specific preferences in their choice.

1(1371). **Micropterix tunbergella** (Fabr.) (*thunbergella* sensu auctt.)
O. 5-6.
L.
P.
I. 5-6. Feeds on the pollen of *Quercus* spp., *Acer pseudoplatanus* and *Crataegus*.

2(1372). **M. mansuetella** (Zell.)
O. 5-6.
L.
P.
I. 5-6. Feeds on the pollen of various herbaceous plants, especially *Carex* spp. growing in very wet situations.

3(1373). **M. aureatella** (Scop)
O. 5-6.
L.
P.
I. 5-6. Feeds on the pollen of various herbaceous plants, especially *Carex* spp. growing on acid soils.

4(1374). **M. aruncella** (Scop.)
O. 6-8.
L.
P.
I. 5-8. Feeds on the pollen of many herbaceous plants and *Crataegus*.

5(1375). **M. calthella** (Linn.)
O. 5-6.
L.
P.
I. 5-6. Feeds on the pollen of many herbaceous plants.

ERIOCRANIIDAE

The larvae feed in blotch mines characterised by the long, intertwining threads of frass. In the birch-feeding species, the mines of nos 9 and 10 start with a gallery; the others have no initial gallery and are hard to distinguish except by larval characters. All have the pupa subterranean, in a tough, silken cocoon. The imagines fly in sunshine, sometimes in swarms round their food-plant; in dull weather they sit on the twigs.

6(1370). **Eriocrania subpurpurella** (Haw.)
O. 5. In a leaf-bud of *Quercus* spp.
L. 5-7.
P. 7-4.
I. 4-5. Comes to light.

7(1364). **E. chrysolepidella** (Zell.) (*kaltenbachii* (Wood))
O. 4. In a leaf-bud of *Corylus* or *Carpinus*.
L. 4-5. Larva whitish.
P. 5-4.
I. 4.

8(1369). **E. unimaculella** (Zett.)
O. 4. In a leaf-bud of *Betula* spp.
L. 4-5. Larva whitish with 2 blackish spots on thoracic segment 1 and small lateral projections on abdominal segment 2; head brown.
P. 5-3.
I. 3-4.

9(1365, 1368). **E. sparrmannella** (Bosc) (*fimbriata* Wals.)
O. 5-6. In a leaf of *Betula* spp.
L. 6-8. Mine initially linear, leading to a round or squarish blotch, generally browner than that of the next species. Larva whitish, thoracic segment 1 with 2 brown spots; head brown.
P. 8-4.
I. 4-5.

10(1367). **E. salopiella** (Staint.)
O. 5. In a leaf of *Betula* spp.
L. 5-6. Mine initially linear, leading to an elongate blotch sometimes even resembling a very broad gallery, generally greener than that of the preceding species. Larva whitish, anterior edge of thoracic segment 1 brownish; head pale brown.
P. 6-4.
I. 4-5.

11(1366). **E. haworthi** Bradley (*purpurella* (Haw.))
O. 4. In a leaf-bud of *Betula* spp.
L. 5. Larva whitish, abdominal segments 3-8 with small lateral projections; head in last instar whitish, mouth dark brown.
P. 5-4.
I. 4.

12(1363). **E. sangii** (Wood)
O. 4. In a leaf-bud of *Betula* spp.
L. 4-5. Larva slate-grey.
P. 5-3.
I. 3-4.

13(1362). **E. semipurpurella** (Steph.)

O. 4. In a leaf-bud of *Betula* spp.

L. 4-5. Larva whitish with small lateral projections on abdominal segment 1; no blackish spots on thoracic segment 1; head pale ochreous brown.

P. 5-3.

I. 3-4.

NEPTICULIDAE

14(1361). **Bohemannia quadrimaculella** (Boh.)

O.

L. Possibly in the bark of shoots of *Alnus glutinosa*.

P.

I. 7-8. Flies in afternoon sunshine or may be obtained by tapping branches of alder over a net.

15(1347). **Etainia decentella** (Herr.-Schäff.)

O. 6;8. In June, on the wing of a samara of *Acer pseudoplatanus*.

L. 7;9-?10. The July larvae feed in the seeds.

P. 5;7. In a cocoon spun on the trunk of the host tree.

I. 6;8. Often rests on a trunk and also comes to light. The life history of the winter generation is unknown.

16(1346). **E. sericopeza** (Zell.) (*Ent.Rec.* **89**: 257-264 (1977)

O. 5-6;8-9. In the spring on a samara of *Acer platanoides*; in the summer at the base of a leaf-stalk.

L. 6;9-5. The June larvae feed in the seeds; the second generation at first mine the base of a petiole and later a flower-bud.

P. 5;7. In July, in a cocoon on the wing of a samara, which generally falls from the tree; in May on the surface of the mined flower-bud. Cocoons of both generations may also be found on trunks.

I. 5;8. Often rests on a trunk.

17(1346A). **E. sphendamni** (Her.)

O. 6;8;10. On the wing of a samara of *Acer campestre*.

L. 6;8;10-?. In a gallery leading into the seed, which it then devours.

P. 5;7;9. In a cocoon spun in May on a trunk and in the summer generations on the wing of a samara or the surface of a leaf.

I. 5-6;7-8;9-10. In three generations in favourable seasons. The life history of the winter generation of larvae is imperfectly known.

18(1354). **Ectoedemia argyropeza** (Zell.)

O. 6. On the stalk of a leaf of *Populus tremula*, about 10mm below the blade.

L. 7-11. Mines up the leaf-stalk and makes a blotch at the base of the leaf; larvae may be found in fallen leaves within conspicuous "green islands".

P. 4. Larvae overwinter in their cocoons and pupate in the spring.

I. 5-6. A parthenogenetic species.

19(1357A). **E. turbidella** (Zell.) (*marionella* (Ford))

O. 6. On the side of a stalk of a leaf of *Populus canescens*, about 10mm below the blade; often one on each side of the petiole.

L. 7-10. Mine similar to that of *E. argyropeza*.

P. 11-5. In a cocoon on the surface of the soil.

I. 5-6. Rests on the boles of the host and neighbouring trees.

20(1344). **E. intimella** (Zell.)

O. 6-7. Laid on the midrib of a leaf of *Salix* spp., especially *S. caprea*.

L. 7-11. At first mines the midrib, then makes a blotch in the leaf with a double line of frass edging the passage from the midrib.

P. 11-6. On the ground or on fallen leaves.

I. 6. May be beaten from branches of the foodplant.

21(1330). **E. agrimoniae** (Frey) (*agrimoniella* (Herr.-Schäff.))

O. 6-7. Laid on the underside of a leaf of *Agrimonia*.

L. 8-11. In a slender, tortuous gallery leading to a large, irregular blotch; often several larvae occupy the same blotch. Larva, which mines venter uppermost, with dark spots on thoracic segments.

P. 11-6. Cocoon in the mine, generally near the tip of a lobe.

I. 6.

22(–). **E. spinosella** (Joan.)

O. 6-7. On the underside of a leaf of *Prunus spinosa*.

L. 7-10. In a contorted gallery with dispersed frass, leading to a small irregular blotch. Larva, which mines venter uppermost, with dark ventral spots.

P. 10-6. In a cocoon spun on detritus or the surface of the soil.

I. 6.

23(1332). **E. angulifasciella** (Staint.)

O. 7-8. On the underside of a leaf of *Rosa* spp. or, occasionally, *Poterium sanguisorba*, normally near the midrib.

L. 9-11. In a contorted gallery with dispersed frass leading to a blotch; the mined area frequently turns purple. Larva, which mines venter upwards, with abdominal dark spots and reddish brown head.

P. 11-6. In a cocoon spun on a leaf-litter.

I. 7.

24(1331). **E. atricollis** (Staint.)

O. 6-7. On the underside of a leaf of *Crataegus* spp., *Malus* spp. or, less frequently, *Pyrus* or *Prunus avium*.

L. 8-4. In a gallery which often follows the leaf-margin, containing dispersed frass and leading to a blotch in which the frass is scattered erratically; exit-hole on the upperside. Larva, which mines venter upwards, with abdominal dark spots, black head and black prothoracic plate. Overwinters full-fed.

P. 4-5. In a cocoon spun on or just under the surface of the soil.

I. 6.

25(1334). E. **arcuatella** (Herr.-Schäff.)
- O. 7-8. On the underside of a leaf of *Fragaria vesca* or *Potentilla sterilis*.
- L. 8-10. In a contorted gallery leading to a large blotch. Larva, which mines venter upwards, with obscure brown ventral spots.
- P. 10-6. In a cocoon spun on or under the surface of the soil.
- I. 6-7.

26(1333). E. **rubivora** (Wocke)
- O. 7-9. On the underside of a leaf of *Rubus* spp., especially *R. caesius*.
- L. 10. In a contorted gallery with dispersed frass leading to a blotch; the mined area sometimes turns dull purple.
- P. 10-6. In a cocoon spun on or below the soil surface.
- I. 6-7.

27(–). E. **erythrogenella** (Joan.)
- O. 7-9. On or beside a rib on the upperside of a leaf of *Rubus fruticosus*.
- L. 10-12. In a gallery following a vein, usually turning back on a contiguous course and completely filled with frass; this leads to a blotch. The mined area turns bright purplish crimson.
- P. 12-7. In a cocoon, often spun on the surface of a leaf.
- I. 7. Apparently confined to coastal localities.

28(–). E. **bradfordi** Emmet
- O.
- L.
- P.
- I. 6. Life history unknown.

29(1322). E. **argentipedella** (Zell.)
- O. 7. On the underside of a leaf of *Betula*.
- L. 8-10. In a round, black-brown blotch which later has pale edges.
- P. 10-6. In a cocoon spun on the surface of the soil.
- I. 6-7.

30(1321). E. **mediofasciella** (Haw.). (*woolhopiella* (Staint.))
- O. 6. On the underside of a leaf of *Betula* or, less usually, *Corylus*.
- L. 7-9. In a highly contorted gallery, staining the leaf brown, which leads to a large blotch accommodating the early gallery in one corner.
- P. 9-5. In a cocoon spun beneath the surface of the soil.
- I. 5-6.

31(1348). E. **quinquella** (Bed.)
- O. 7-10. On the underside of a leaf of *Quercus robur* or, possibly, other deciduous species of *Quercus*.
- L. 10-11. In a slender, contorted gallery situated in a "green island", sometimes a large number in a leaf. Larva in early instars with large dark spots on the venter. Larvae are best found in fallen leaves.
- P. 11-6. In a cocoon spun amongst detritus.
- I. 6-7. Often rests on the trunk.

32(1352). **E. albifasciella** (Hein.)

O. 7. On the upperside of a leaf of *Quercus* spp., beside a rib.

L. 8-9. In a fine gallery, following a vein leading to a squarish blotch generally away from the midrib. Larva with head and prothoracic plate very pale brown.

P. 9-6. In a cocoon spun on detritus.

I. 6. Often rests on a trunk.

33(1351). **E. subbimaculella** (Haw.)

O. 7-10. On the upperside of a leaf of *Quercus* spp., beside a rib.

L. 10-11. In a fine gallery, usually leading to the angle made by a vein and the midrib, where the larva constructs a blotch with a slit in the lower epidermis of the leaf, allowing much of the frass to fall out. Larva with head and prothoracic plate blackish brown.

P. 11-6. In a cocoon spun on detritus.

I. 6-7. Often rests on a trunk.

34(–). **E. heringi** (Toll). (*quercifoliae* (Toll)

O. 7-10. On the upperside of a leaf of *Quercus* spp., beside a rib.

L. 10-11. In a mine similar to that of the preceding species, but without the slit in the epidermis. Larva with head and prothoracic plate reddish brown.

P. 11-6. In a cocoon spun on detritus.

I. 6-7. Often rests on a trunk.

35(1357). **E. pulverosella** (Staint.)

O. 5-6. On either side of a leaf of *Malus* or *Pyrus*.

L. Mid 6-early 7. In a gallery, often following the leaf-margin and leading to a large, brownish blotch; exit-hole on the underside. Larva yellow with pale brown head.

P. 7-5. In a cocoon spun on detritus.

I. 5.

36(1358). **E. atrifrontella** (Staint.)

O. On the young green bark of *Quercus* spp.

L. In a tortuous gallery in the bark; vertical trunks up to the thickness of one's arm in full sunshine are preferred.

P.

I. 6-8. May be beaten from the foodplant.

37(1342). **Fomoria septembrella** (Staint.)

O. 6;8-9. On the underside of a leaf of *Hypericum* spp., especially *H. perforatum*.

L. 7;10-12. In a fine gallery leading to a large blotch in the centre of the leaf.

P. 7-8;12-5. In a cocoon spun in an inflated cavity in the centre of the blotch.

I. 5-6;8.

38(1345). **F. weaveri** (Staint.)

O. 6-8. On the underside of a leaf of *Vaccinium vitis-idaea*.

L. 8-5. In a fine gallery leading to a large blotch in the centre of the leaf.

P. 5-8. In a cocoon spun in an inflated cavity in the centre of the blotch; leaves containing cocoons tend to fall off the plant.

I. End 5-8. The emergence of adults takes place over an extended period.

39(1353). **Fedalmia headleyella** (Staint.)

O. 7-9. On the upperside of a leaf of *Prunella vulgaris*, usually on a seedling in long grass.

L. 7-10. In a fine gallery leading into the petiole and thence, via the stem, into a second leaf where it makes a blotch; sometimes three leaves are mined. The mined area tends to turn purple.

P. 10-5 or later. In a cocoon spun on detritus or the upper surface of a leaf of the foodplant.

I. End 5-8. There is probably only one generation with a long emergence period.

40(–). **Trifurcula griseella** Wolff

O.

L.

P.

I. 5;8. Life history unknown; occurs on downland.

41(1359). **T. immundella** (Zell.)

O. 8-9. On the bark of a twig of *Sarothamnus*, near the tip. The egg remains conspicuous throughout the mining period.

L. 10-5. In a gallery in the bark; the larva starts by mining downwards.

P. 6-7. In a cocoon spun on detritus.

I. 7-8. Obtained by tapping branches of the foodplant over a net; also comes to light.

42(1360). **T. pallidella** Zell.

O.

L.

P.

I. 8-9. Life history unknown; possibly associated with *Genista tinctoria*. Flies for a short period at dusk. Has occasionally been taken in early July.

43(1356). **T. cryptella** (Staint.)

O. 6. On the underside of a leaf of *Lotus*, preferring *L. uliginosus* growing in damp or shady situations.

L. 6-8 or even later. A fine gallery, often following the leaf margin, leading to a blotch which generally occupies the whole leaflet.

P. 8-5. In a cocoon spun on detritus or the surface of the soil.

I. 5-6.

44(1355). **T. eurema** (Tutt)

O. 6;9. On either side of a leaf of *Lotus*, preferring *L. corniculatus* growing on downland.

L. 6-7;9-10. Mine similar to that of *T. cryptella*.

P. 7;10-5. In a cocoon spun on detritus or the surface of the soil.

I. 5-6;8. The habitat is generally a reliable means of distinguishing this species from *T. cryptella;* there is no constant difference between the mines.

45(1295). **Stigmella aurella** (Fabr.) (? (1297)*gei* (Wocke) partim)

O. 4;7;9.	On either side of a leaf of *Rubus* spp., preferring *R. fruticosus*; also, probably, on *Geum*.
L. 5-6;7-8; 10-3.	In a long gallery containing dispersed frass. The winter generation overwinters half-fed in the mine.
P. 3-4;6;8.	In a cocoon spun on detritus.
I. 4;7;9.	The timing and number of generations depends on the severity of the winter.

46(1296). **S. fragariella** (Heyd.) (?*nitens* (Fologne))

O. 4;8.	On the upperside (occasionally the underside) of a leaf of *Agrimonia, Fragaria vesca, Rubus caesius* or *R. idaeus*.
L. 6-7;8-10.	In a mine similar to that of *S. aurella; S. nitens* which has the frass linear and not dispersed, is probably only a mine-form.
P. 7;10-4.	In a cocoon spun on detritus.
I. 4;8.	Emerges with the first warm weather in spring, the timing of the subsequent generation, therefore, being variable. Probably distinct from *S. aurella* because of the different way of passing the winer. Male adults usually have the head black.

47(1298). **S. dulcella** (Hein.)

O. 5-6;8-9.	On the upperside of a leaf of *Fragaria vesca* and probably also *Rubus fruticosus*.
L. 7;9-10.	In a long gallery in which the frass is linear or smudged, only seldom dispersed; usually there are several larvae in a leaf.
P. 7-8;10-5.	In a cocoon spun on detritus.
I. 5;8.	Probably distinct from *S. fragariella* because of its smaller size and duller colouration; in the male the head is always ferruginous.

48(1294). **S. auromarginella** (Rich.)

O.	Laid on the upperside of a leaf of *Rubus fruticosus* or *R. saxatilis*.
L.	Mine similar to that of *S. aurella*.
P.	In a cocoon spun on detritus.
I.	More or less continuously brooded, emerging even in mid-winter if the weather is mild; adults are more frequent in May and August-September.

49(1293). **S. splendidissimella** (Herr-Schäff.)(? (1227) *gei* Wocke, partim)

O. 5-6;8-9.	Usually on the upperside of a leaf on *Rubus caesius* or, less often, other species of *Rubus*, generally in shady situations; also on *Geum*. On the mountains of Scotland on *Rubus chamaemorus* and *R. saxatilis*.
L. 7;9-10.	In a long, sinuous gallery with linear frass.
P. 7-8;10-4.	In a cocoon spun on detritus.
I. 4-5;8.	Univoltine on high ground in Scotland.

50(1292). **S. aeneofasciella** (Herr.-Schäff.)

O. 5-6;8-9.	Laid usually on the underside of a leaf of *Agrimonia, Potentilla erecta, P. reptans, P. anserina* or *Fragaria vesca*.

L. 7;9-11. In a slender gallery leading to a blotch. Larva yellow without dark spots on the ventral surface of the thoracic segments.
P. 7-8;11-5. In a cocoon spun on detritus.
I. 5;8.

51(–). **S. dryadella** (Hofm.)
O. 6;9. On the underside close to the margin, or even on the extreme edge, of a leaf of *Dryas*.
L. 7;9-10. In a gallery usually following the leaf-margin, leading to a yellowish brown blotch.
P. 7-8; 10-5. In a cocoon spun on detritus or occasionally at the base of a leaf-stalk.
I. 5-6;8-9.

52(1302). **S. filipendulae** (Wocke)
O. 6;8. On the upperside of a leaf of *Filipendula vulgaris*.
L. 7; late 8-10. In a rather broad gallery following the leaf-margin leading to a blotch which often occupies the whole area between the margin and the midrib.
P. 7-8;10-5. In a cocoon spun on detritus.
I. 5;8.

53(–). **S. ulmariae** (Wocke)
O. 5-6;8-9. On a leaf of *Filipendula ulmaria*, generally on the upperside.
L. 6-7;9-10. In an irregular gallery.
P. 7-8;10-5. In a cocoon spun on detritus.
I. 5;8.

54(1300). **S. poterii** (Staint.)
O. 5;8-9. On the upperside of a leaf of *Poterium sanguisorba*.
L. 6-7;9-10. In a slender gallery following the leaf-margin, widening into a blotch occupying nearly the whole of the leaflet.
P. 7-8;10-4. In a cocoon spun on detritus.
I. 4-5;8.

55(1299). **S. tengstroemi** (Nolck.)
O. 6. On either side of a leaf of *Rubus chamaemorus*.
L. 7. In an irregular gallery.
P. 8-5. In a cocoon spun on detritus.
I. 5-6. Occurs at about 600m in the Scottish Highlands. Some adults emerge in the August of the year in which the larva fed, but it is doubtful whether their offspring can complete their cycle before the foodplant dies down.

56(1301). **S. serella** (Staint.)
O. 6-7;8-9. On a leaflet of *Potentilla erecta*, usually on the upperside. Plants growing in long grass or heather are preferred.
L. 7;9-10. In a slender gallery following the leaf-margin leading to a blotch which may occupy the whole of the leaflet.
P. 7-8;10-6. In a cocoon spun on detritus.
I. 6;8. Supposed British specimens of *S. tormentillella* (Herr.-Schäff.) are referable to this species.

57(1307). **S. marginicolella** (Staint.)
O. 6;8-9. On either side of a leaf of *Ulmus* spp.
L. 7;9-10. In a long, sinuous gallery, sometimes following the leaf-

margin. The frass in the early part of the mine is dispersed, almost filling the gallery; exit-hole on the upperside. Larva yellow.

P. 7-8;10-5. In a cocoon spun on detritus.

I. 5;8.

58(1310). **S, continuella** (Staint.)

O. 5-6;8-9. On the underside of a leaf of *Betula*, often in an angle of veins.

L. 6-7;9-10. In a gallery starting from a brownish blotch and in its later course completely filled with greenish frass; larva rich yellow but appearing green in the mine.

P. 7;10-5. In a cocoon spun on detritus.

I. 5-6;8. Some June larvae produce adults in August; others the following May.

59(1308). **S. speciosa** (Frey)

O. 6;8. On the underside of a leaf of *Acer pseudoplatanus*.

L. 7-8;9-10. In a long irregular gallery containing dispersed or linear frass. Larva yellow.

P. 8;10-6. In a cocoon spun on detritus.

I. 6;8. Most July larvae do not produce imagines till the following year, but a few emerge in August and produce the second generation.

60(1320). **S. sorbi** (Staint.)

O. 5. On the underside of a leaf of *Sorbus aucuparia*, occasionally on other *Sorbus* spp., or *Cotoneaster simonsii*.

L. 6. In a slender gallery leading abruptly to a large, oval blotch.

P. 6-5. In a cocoon spun on detritus.

I. 5. The stages occur up to a month later in the north.

61(1315). **S. plagicolella** (Staint.)

O. 6;8. On the underside of a leaf of *Prunus spinosa* or, less frequently, other *Prunus* spp.

L. 7;9-10. In a slender gallery leading abruptly to a round blotch; in the autumn generation, the blotch is more elongate.

P. 7-8;10-5. In a cocoon spun on detritus.

I. 5-6;8.

62(1336). **S. salicis** (Staint.)

O. 5;8-9. On the underside of a leaf of one of the rough-leaved *Salix* spp. (*S. caprea, S. cinerea* or *S. aurita*).

L. 6-7;9-11. In a contorted gallery, often leading to a small blotch.

P. 7-8;11-4. In a cocoon spun on detritus.

I. 4-5;7-8.

63(–). **S. auritella** (Skala) *Ent. Rec.* **88**:105-109(1976)

O. 5;8-9. On the underside of a leaf of *Salix aurita*.

L. 6-7;9-10. In a mine similar to that of *S. salicis*, but smaller and with a reddish tinge.

P. 7-8;10-4. In a cocoon spun on detritus.

I. 4-5;7-8.

64(1338). **S. obliquella** (Hein.)

O. 5-6;8-9. On either side of a leaf of the smooth-leaved *Salix* spp. (*S. fragilis, S. alba, S. viminalis, S. babylonica, S. purpurea, S. petandra*).

L. 6-7;9-10. In a mine similar to that of *S. salicis*.

P. 7-8;10-5. In a cocoon spun on detritus or, in the summer generation, on a leaf of the food-plant.

I. 5;8. There is no evidence for the occurrence of (1337) *S. vimineticola* (Frey) in Britain; it was formerly confused with *S. obliquella*.

65(–). **S. repentiella** (Wolff) *Ent. Rec.* **89**:178-182 (1977)

O. 5-6;8-9. On the underside of a leaf of *Salix arenaria*.

L. 7;9-10. In a gallery leading to the leaf-margin, which it follows before widening into a small blotch.

P. 7-8;10-5. In a cocoon spun on detritus.

I. 5;8. On coastal sandhills.

66(1335). **S. myrtillella** (Staint.)

O. 6;8. On the underside of a leaf of *Vaccinium myrtillus* or *V. uliginosum*, generally close to the midrib.

L. 7;9-10. In a contorted gallery leading to an irregular blotch; mined leaves tend to turn purple.

P. 7-8;10-5. In a cocoon spun on detritus.

I. 5-6;8.

67(1349). **S. trimaculella** (Haw.)

O. 5-6;8-9. On either side of a leaf of *Populus nigra* including var. *italica*, or *P. x canadensis*.

L. End 6-7; 9-10. In a gallery containing either linear or dispersed frass. Mines may be partly or wholly on the underside of the leaf.

P. 7;10-5. In a cocoon spun on detritus.

I. 5;8. Often rests on a trunk.

68(1350). **S. assimilella** (Zell.)

O. 7. On the upperside of a leaf of *Populus tremula*, generally beside a rib; less often on *P. alba*.

L. 8-10. In an irregular gallery, sometimes leading to a small blotch.

P. 10-6. In a cocoon spun on detritus.

I. 6-7. There is probably a single generation with a long emergence period.

69(1339). **S. floslactella** (Haw.)

O. 5-6;8-9. On the underside of a leaf of *Corylus* or *Carpinus*.

L. 6-7;9-10. In a gallery of greater width than that of the larva; the early part is filled, or almost filled, with dispersed frass. Exit-hole on the upperside. Larva duller yellow than *S. microtheriella*; mines dorsum upwards.

P. 7;10-5. In a white, fluffy cocoon spun on detritus.

I. 5;8.

70(–). **S. carpinella** (Hein.) *Ent. Rec.* **99**:64-66(1977)

O. 5-6;8-9. On the underside of a leaf of *Carpinus*.

L. 6-7;9-10. In a gallery similar to that of *S. floslactella*.

P. 7;10-5. In a yellowish white cocoon, rather less fluffy than that of *S. floslactella*, spun on detritus.

I. 5;8. Rests on a trunk.

71(1325,1327). **S. tityrella** (Staint.) (*turicella* (Herr.-Schäff.))

O. 5-6;8. On the underside of a leaf of *Fagus*, generally under the fluffy hair growing in the angle between the midrib and a lateral vein.

L. 6-7; end8-10. In a sinuous gallery between veins or near the leaf-margin; frass never in arcs. Larva yellow.

P. 7-8;10-4. In a cocoon spun on detritus.

I. End4-5;7-8. Often rests on a trunk.

72(1274). **S. pomella** (Vaugh.)

O. 5-6;9. On a leaf of *Malus*, usually on the underside close to a rib.

L. 6-7;10. In a narrow gallery leading to a small orange-brown blotch alongside a vein or in an angle between veins. Larva yellow with pale brown head.

P. 7-8;10-5. In a cocoon spun on detritus.

I. 5;8.

73(1275). **S. perpygmaeella** (Doubl.) (*pygmaeella* (Haw.))

O. 6;9. On a leaf of *Crataegus*, usually on the upperside beside the midrib; difficult to find.

L. 7;10. In a fine gallery leading to a series of "S" turns fusing to form a small blotch, all more compact than the mine of *S. crataegella*. Larva yellow.

P. 7-8;10-5. In a cocoon spun on detritus.

I. 5;8.

74(1305). **S. ulmivora** (Fologne)

O. 6. On the underside of a leaf of *Ulmus* generally close to a rib.

L. 8-9. In a very variable gallery; in its early part the frass is always linear leaving clear margins; thereafter it may stay linear or be dispersed or coiled; exit-hole on the underside. Larva green.

P. 9-5. In a cocoon spun on detritus.

I. 5. In forward seasons some larvae feed up quickly and give rise to a small second generation. The supposed species *S. ulmifoliae* (Hering) and *S. ulmicola* (Hering) are variant mine-forms. Another mine-form has been mistaken for *S. ulmiphaga* (Preissecker), which does not occur in Britain.

75(1324,1326). **S. hemargyrella** (Kollar) (*basalella* (Herr.-Schäff.); *fulgens* (Staint.))

O. 5;8. On either side of a leaf of *Fagus*.

L. 6;8-early 9. In an irregular gallery, in the middle portion of which the frass is arranged in a series of arcs. Larva pale yellowish white.

P. 6;9-4. In a cocoon spun on detritus.

I End 4-5; end 6-early 8. Rests on trunks.

76(–). **S. paradoxa** (Frey)

O. 5-6. On the underside of a leaf of *Crataegus*, at the extreme tip of a lobe.

L. 6-early 7. In a round blotch with the frass in the centre.

P. 7-4. In a cocoon spun on detritus.

I End 4-5. Rests on a trunk by day and flies in the late afternoon.

77(1276). **S. atricapitella** (Haw.)
O. 6;8-9. On the underside of a leaf of *Quercus* spp., sometimes close to the margin.
L. 6-7;9-11. In an irregular gallery which, in the first generation, is relatively short with a thick line of frass; in the second generation the gallery is longer and the frass-line finer.
P. 7-8;10-5. In a cocoon spun on detritus.
I. 5;7-8. Often rests on a trunk.

78(1277,1328). **S. ruficapitella** (Haw.) (?*castanella* (Staint.))
O. 6;8-9. On the upperside of a leaf of *Quercus* spp.
L. 6-7;9-10. In an irregular gallery with dispersed frass in its second half; feeds about a fortnight earlier than related species.
P. 7;10-5. In a cocoon spun on detritus.
I. 5;7-8. Often rests on a trunk.

79(1287). **S. suberivora** (Staint.)
O. 5-6;9. On the upperside of a leaf of *Quercus ilex*.
L. 7;10-3. In a broad, irregular gallery almost filled with black frass.
P. 4;7-8. In a cocoon spun on detritus.
I. 4-5;8.

80(–). **S. roborella** (Johansson)
O. 6;8-9. On the underside of a leaf of *Quercus* spp.
L. End 6-7; 10-11. In the summer generation, in rather a small mine with a relatively thin line of frass; that of the autumn generation resembles the mine of *S. atricapitella*.
P. 7-8;11-5. In a cocoon spun on detritus.
I. 5-6;8.

81(–). **S. svenssoni** (Johansson)
O. 6-7. On the underside of a leaf of *Quercus* spp., beside a vein.
L. 7-9. In a long irregular gallery, leading at first directly away from the vein, then extending well across the leaf, the mine being the longest and broadest of those of the oak-feeding species; in the second half of the mine, the frass is dispersed.
P. 8-5. In a cocoon spun upon detritus.
I. 5-6. There is one extended generation.

82(–). **S. samiatella** (Zell.) *Ent. Rec.* **88**:315-318 (1976)
O. 6;8-9. On the underside of a leaf of *Quercus* spp., beside a vein.
L. 6-7;9-10. In a gallery similar to that of *S. svenssoni*, but slightly smaller.
P. 7-8;10-5. In a cocoon spun on detritus.
I. 5;8.

83(1278). **S. basiguttella** (Hein.)
O. 6;8-9. On the upperside of a leaf of *Quercus* spp.
L. 6-7;9-10. In a long, sinuous gallery entirely filled with greenish frass. Larva green.
P. 7;10-5. In a cocoon spun on detritus.
I. 5-6;7-8.

84(1279). **S. tiliae** (Frey)
O. 6;8-9. On the underside of a leaf of *Tilia cordata* (occasionally *T. x vulgaris*), generally close to a rib.

L. 7;9-10. In a highly contorted gallery, at first confined to the underside of the leaf.
P. 7-8;10-5. In a cocoon spun on detritus or the surface of a leaf.
I. 5-6;8.

85(1284). **S. minusculella** (Herr.-Schäff.)
O. 5-6;8-9. On the underside of a leaf of *Pyrus*.
L. 6-7;8-9. In an irregular gallery, rather broad in its second half, with linear frass. Larva green.
P. 7-8;9-5. According to some accounts, the autumn larvae overwinter and do not pupate until the spring.
I. 5;8.

86(1280,1281). **S. anomalella** (Goeze) (*fletcheri* (Tutt))
O. 6;8-9. In a leaf of *Rosa* spp. (including *R. pimpinellifolia*), generally on the underside.
L. 7;10-12. In a long sinuous gallery, the early part of which is generally completely filled with frass. Later the frass may be dispersed in arcs, or linear. Late larvae may go on feeding until December.
P. 7-8;10-5. In a cocoon spun, in the summer brood, at the base of a leaf-stalk or on a leaf.
I. 5;8.

87(1311,1312). **S. centifoliella** (Zell.) (*hodgkinsoni* (Staint.))
O. 5-6;8-9. On a leaf of *Rosa* spp., generally on the underside.
L. 6-7;9-10. In a long, sinuous gallery with linear frass; the mine is indistinguishable from that of *S. anomalella* when that has linear frass.
P. 7;10-4. Pupation habits similar to those of *S. anomalella*.
I. 4-5;7-8.

88(1323). **S. spinosissimae** (Waters)
O. 6;8-9. On a leaf of *Rosa pimpinellifolia*, generally on the underside near the midrib.
L. 7;9. In a gallery resembling that of *S. anomalella*.
P. 7-8;10-5. Pupation habits similar to those of *S. anomalella*.
I. 5;8. The early stages are indistinguishable from those of *S. anomalella*.

89(1282). **S. viscerella** (Staint.)
O. 6-7. On the upperside of a leaf of *Ulmus*.
L. 8-9. In a gallery progressing in a series of contiguous "S" turns, so as to give the appearance of a blotch; exit-hole on the underside. Larva green.
P. 10-5. In a cocoon spun on detritus.
I. End 5-6.

90(1329). **S. malella** (Staint.)
O. 5-6;8-9. On the underside of a leaf of *Malus*, generally near a vein.
L. 6-7;9-10. In an irregular gallery which widens conspicuously, containing erratically deposited linear frass. Larva yellow with brown head.
P. 7-8;10-5. In a cocoon spun on detritus.
I. 5;8.

91(1343). **S. catharticella** (Staint.)

O. 6;8-9. On the underside of a leaf of *Rhamnus*, usually close to the midrib.

L. 6-7;9-10. In a gallery progressing in a series of "S" turns and more or less filled with greenish frass.

P. 7;10-5. In a cocoon spun on detritus.

I. 5-6;7-8.

92(1316). **S. hybnerella** (Hübn.) (*ignobilella* (Staint.))

O. 5;early 8. On a leaf of *Crataegus*, nearly always on the underside near a vein, well away from the leaf margin.

L. End 5-6; 8-9. In a fine gallery with linear frass, widening abruptly into a blotch, usually near the edge of the leaf. Larva whitish with brown head and without a dark prothoracic plate.

P. 7;9-4. In a cocoon spun on detritus.

I. 4-5;7-8. In favourable seasons a third generation occurs in the autumn.

93(1286). **S. oxyacanthella** (Staint.)

O. 7-8. On the underside of a leaf of *Crataegus, Malus, Pyrus, Cotoneaster frigidus* or, possibly, *Sorbus aucuparia*.

L. 9-10. In a long gallery, which at first follows a vein and contains linear frass, and later traverses the leaf, when the frass is dispersed or in arcs. Larva green.

P. 10-6. In a cocoon spun on detritus or the surface of the soil.

I. 6. When feeding on apple, has been mistaken for *S. aeneella* (Hein.) which does not occur in Britain.

94(1285). **S. pyri** (Glitz)

O. 6;8-9. On a leaf of *Pyrus* usually on the upperside.

L. 7;9-early 10. In a narrow, strongly contorted gallery containing black, dispersed frass. Larva bluish green with the gut red.

P. 7-8;10-5. In a cocoon spun on detritus.

I. End 5-6;8.

95(–). **S. aceris** (Frey) *Ent. Rec.* **89**:244-247 (1977)

O. 6;8. On the underside of a leaf of *Acer campestre, A. platanoides* or, less usually, *A. pseudoplatanus*.

L. End 6-early 7;8-early 9. In a long, sinuous gallery filled with green frass. Larva green.

P. 7;9-5. In a white cocoon spun on the surface of a leaf or on detritus.

I. 5;7-8.

96(1283). **S. nylandriella** (Tengst.) (*aucupariae* (Frey))

O. 6. On the underside of a leaf of *Sorbus aucuparia*.

L. 7-8. A long sinuous gallery, sometimes following the margin of the leaf; the frass is usually well dispersed but may, occasionally, be linear.

P. 8-5. In a cocoon spun on detritus.

I. 5-6. There is possibly a second generation in favourable seasons.

97(1289). **S. magdalenae** (Klimesch) (*nylandriella* sensu auctt.)

O. 6. On the underside of a leaf of *Sorbus aucuparia*.

L. 7-8. In a small, contorted gallery, sometimes fusing to form a small blotch, with fine, linear frass.

P. 8-5. In a cocoon spun on detritus.
I. 5-6.

98(1288). **S. desperatella** (Frey)
O. 7. On the underside of a leaf of *Malus sylvestris*, generally one of the small, low leaves of saplings.
L. End 8-early 10. In a slender gallery, widening into a gallery so contorted as to form a blotch; frass linear. Larva bluish green.
P. 10-6.
I. 6.

99(1290). **S. torminalis** (Wood)
O. 6. On the underside of a leaf of *Sorbus torminalis*.
L. 7. In a slender gallery, widening almost into a blotch; frass linear.
P. 8-5.
I. 5-6.

100(1291). **S. regiella** (Herr.-Schäff.)
O. 7. On the underside of a leaf of *Crataegus*, close to the margin.
L. 8-early 11. In a narrow gallery, filled with reddish frass, following the leaf margin; this leads abruptly to a pale yellow blotch containing black, irregular frass. Larva rather bright yellow.
P. 11-5. In a cocoon spun on detritus.
I. 5-6. Occasionally in forward seasons there is a small generation earlier in the year.

101(1304). **S. crataegella** (Klimesch) (*gratiosella* sensu Wood)
O. 6. On the underside of a leaf of *Crataegus*, usually near the midrib at the base of the leaf.
L. End 6-early 8. At first a fine gallery, often directed down and back along the petiole; later a broad gallery with dispersed frass, usually forming a blotch by the fusion of "S" turns. Larva green.
P. 8-5. In a cocoon spun on or below the surface of the soil.
I. 5-early 6.

102(1306). **S. prunetorum** (Staint.)
O. 6;8-9. On the underside of a leaf of *Prunus spinosa*, occasionally *P. avium* or *P. cerasus*; generally close to a rib.
L. 7;9-early 10. In a gallery, at first so contorted as to form a small blotch, afterwards in an irregular track, sometimes following the leaf-margin. Larva green.
P. 7-8;10-5. In a cocoon spun on detritus.
I. 5;8.

103(1314). **S. betulicola** (Staint.)
O. 6;8-9. On the underside of a leaf of *Betula* spp., especially on seedlings amongst heather or on boggy ground.
L. 7;9-early 11. In an irregular gallery, only slightly contorted at the start and with irregular, linear frass. Larva, which mines venter upwards, with a chain of dark ventral spots, and brown head.
P. 7-8;10-5. In a cocoon spun on detritus.
I. 5;8.

104(–). **S. betulicola nanivora** (Petersen)

O. 6;8-9. On *Betula nana*, on the underside of a leaf.

L. 7;9-10. Life history similar to that of *S. betulicola*. Univoltine in unfavourable seasons, when autumn larvae only occur.

P. 7-8;10-5.

I. 5;8.

105(1313). **S. microtheriella** (Staint.)

O. 6;8-9. On the underside of a leaf of *Corylus* or *Carpinus*, usually close to a rib.

L. End 6-7; 9-11. In a long, slender gallery, no wider than the width of the larva, frequently following the veins; the early part contains linear frass, leaving clear margins; exit-hole on either side of the leaf. Larva rather bright yellow; mines venter upwards. Often there are many mines in a leaf.

P. 7-8;10-5. In a cocoon spun on detritus.

I. 5;8. A parthenogenetic species.

106(1319). **S. luteella** (Staint.)

O. 6-7. On the underside of a leaf of *Betula* spp.

L. 8-11. In an irregular gallery which at first is much contorted and situated on the underside of the leaf; otherwise similar to that of *S. betulicola*. Larva, which mines dorsum upwards, without ventral spots and with a pale, yellow-brown head.

P. 11-5. In a cocoon spun on detritus.

I. 5-7.

107(1317). **S. distinguenda** (Hein.)

O. 6-8. On a leaf of *Betula* spp., generally on the underside.

L. 7-10. In an irregular gallery with the frass in arcs, leaving only a very narrow clear margin. Larva, which mines venter upwards, with brown head and conspicuous ventral dark spots.

P. 8-5. In a cocoon spun on detritus.

I. 5-6. Apparently in a single, very much protracted, generation.

108(1318). **S. glutinosae** (Staint.)

O. 6;9. On the underside of a leaf of *Alnus glutinosa*, generally beside a rib.

L. 7;9-10. In a long, irregular gallery with the frass usually dispersed but sometimes linear. Larva, which mines venter upwards, usually has the prothoracic plate darker than the abdomen.

P. 7-8;10-5. In a cocoon spun on detritus.

I. 5;8. In many years the earlier generation appears to be absent.

109(1309). **S. alnetella** (Staint.)

O. 6;9. On the underside of a leaf of *Alnus glutinosa*, generally beside a rib.

L. 7;9-10. In a long irregular gallery with the frass linear. Larva, which mines dorsum upwards, without a darker prothoracic plate.

P. 7-8;10-5. In a cocoon spun on detritus.

I. 5-6;8. In many years the earlier generation appears to be absent.

110(1340). **S. lapponica** (Wocke)

O. 5-6.	On the underside of a leaf of *Betula* spp., generally beside a rib.
L. Mid 6-early 7.	In a long, angular gallery, frequently following a vein; the early part is completely filled with cloudy green frass.
P. 7-5.	In a cocoon spun in the soil.
I. 5.	

111(1341). **S. confusella** (Wood)

O. 6.	On the underside of a leaf of *Betula* spp., beside a rib.
L. 7-8.	In a long gallery resembling that of *S. lapponica* except that the black frass is linear throughout.
P. 8-5.	In a cocoon spun in the soil.
I. 5.	In all stages about a fortnight later than *S. lapponica.*

112(1303). **Johanssonia acetosae** (Staint.)

O. 5;7-8.	On the underside of a leaf of *Rumex acetosella* or *R. acetosa*.
L. 5-6;8-10.	In a gallery arranged in a close spiral, tending to turn bright red; sometimes there are many mines in one leaf.
P. 6-7;10-4.	In a cocoon spun on detritus.
I. 4-5;7.	The timing of the stages in variable; in a favourable season there are probably three generations.

OPOSTEGIDAE

113(1156). **Opostega salaciella** (Treits.)

O.	
L.	Possibly in a stem of *Rumex acetosella*.
P.	
I. 6-7.	Rests on the ground or low down on herbage in the day-time. Flies from 1900 - 2000 hrs and later comes to light.

114(1157). **O. auritella** (Hübn.)

O.	
L.	Possibly in a stem of *Caltha palustris*.
P.	
I. 6-7.	In fens.

115(1158). **O. crepusculella** Zell.

O.	
L.	Possibly on *Mentha* spp.
P.	
I. 6-7.	In fens and marshes. Flies at dusk and later comes to light.

116(1159). **O. spatulella** (Herr.-Schäff.)

O.	
L.	
P.	
I. 8-6.	Flies or rests on blades of grass in late afternoon. Hibernates in thatch.

TISCHERIIDAE

The Tischeriidae differ from the majority of lepidopterous leaf-miners in that the frass is ejected from their mines through a hole in the epidermis of the leaf.

117(1171). **Tischeria ekebladella** (Bjerk.) (*complanella* (Hübn.))

O. 7. On the upperside of a leaf of *Quercus* spp. or *Castanea*.

L. 9-4. In a white blotch on the upper surface of the leaf, overwintering full-fed in a circular chamber in the centre of the mine.

P. 5. In the chamber.

I. 6. In very hot seasons there is a summer generation of larvae producing adults in August.

118(1172). **T. dodonaea** (Staint.)

O. 7-8. On the upperside of a leaf of *Quercus* spp. or *Castanea*.

L. 9-4. In a brown blotch with darker, concentric rings on the upper surface of the leaf. Overwinters like the preceding species.

P. 5. In the overwintering chamber.

I. 6.

119(1173). **T. marginea** (Haw.)

O. 6-7;8-9. On the upper surface of a leaf of *Rubus* spp., especially *R. fruticosus*.

L. 7;9-3. In a pyriform, pale brown blotch with an inflated central gallery.

P. 4-5;7-8. In the mine.

I. 5-6;8.

120(1174). **T. gaunacella** (Dup.)

O. Near the midrib on a leaf of *Prunus spinosa* or *P. cerasus*.

L. 9-10. In an inflated blotch on the upperside, causing the leaf to fold over.

P. 10-4. In the mine.

I. 5-6.

121(1175). **T. angusticollella** (Dup.) (*angusticolella* missp.)

O. On a leaf of *Rosa* spp., often close to the ground.

L. 9-10. In an inflated, pale brown blotch on the upperside, causing the leaf to fold over.

P. 11-5. In the mine.

I. 5-6. Occasionally there is a summer generation of larvae, producing adults in August.

INCURVARIIDAE

122(1249). **Phylloporia bistrigella** (Haw.)

O. 6-7. On a leaf of *Betula*, usually near the tip.

L. 7-8. In a narrow gallery terminating in a large blotch.

P. 9-5. In an oval case cut from the blotch, which then falls to the ground.
I. 5-7. Flies round small birches in late afternoon. Occasionally there is a small second generation in the autumn.

123(1250). **Incurvaria pectinea** (Haw.) (*zinckenii* (Zell.))
O. 5-6. On a leaf of *Betula, Corylus, Malus, Carpinus*, usually several on one leaf. *Also recorded on *Alnus, Acer, Prunus, Pyrus, Sorbus* and *Swida*.
L. 6-9. At first in a circular blotch; later in a portable case cut from the blotch, on dead leaves on the ground. Case enlarged with larger pieces cut out from dead leaves.
P. 10-4. In the case attached to a vertical surface, such as the base of a tree-trunk.
I. 4-5. Flies in sunshine from about 0700 - 1100 hrs.

124(1251). **I. masculella** ([D. & S.]) (*muscalella* (Fabr.))
O. 5-6. On a leaf of *Crataegus* or, less often, that of some other tree or shrub.
L. 6-3. Feeding habits similar to those of *I. pectinea*; overwinters full-fed in the case.
P. 3-4. In the case on the ground or occasionally in a crevice at the base of a tree or fence.
I. 4-5. Flies in sunshine.

125(1253). **Lampronia oehlmanniella** (Hübn.)
O. 7. On a leaf of *Vaccinium myrtillus, Rubus chamaemorus, Swida* or *Prunus* spp.
L. 8-3. At first mining a blotch; later in a portable case cut from the blotch, on dead leaves, principally of its foodplant. Case enlarged as in two preceding species.
P. 4-5. In the case attached to an object on the ground.
I. 6-7. Flies in sunshine.

126(1254). **L. praelatella** ([D. & S.])
O. 6-7. On the leaf-margin near the tip of a leaf of *Fragaria* or, less usually, one of the other herbaceous Rosaceae.
L. 7-5. First mining, then in a portable case cut from the mine and later enlarged with silk; it cuts off a segment from the tip of the leaf and attaches it to the underside of the same leaf, feeding on it as it withers. These excised segments betray the presence of the larva.
P. 5-6. In the case.
I. 6. Flies in sunshine.

127(1252). **L. capitella** (Clerck)
O. 6. In green fruits of *Ribes* spp., especially *Ribes rubrum*.
L. 6-4. In a seed of green fruit; after hibernation in a shoot, causing the leaves at its tip to droop.
P. 4-5. Usually in the mined shoot.
I. 5-6. Flies in the afternoon sunshine.

128(–). **L. flavimitrella** (Hübn.) *Ent. Gaz.* **28**:180-182 (1977)

O.

L. Possibly associated with *Rubus idaeus* and/or other *Rubus* spp.

P.

I. 5. Flies in sunshine at about 1100 hrs.

129(1255). **L. luzella** (Hübn.)

O.

L. Possibly associated with *Rubus* spp.

P.

I. 5-7. In woods. Flies in the afternoon sunshine.

130(1256). **L. rubiella** (Bjerk.)

O. In the receptacle of a flower of *Rubus idaeus* or other spp. of *Rubus*.

L. 7-5. When young, in the fruit receptacles; after hibernation, in developing buds and adjacent pith.

P. 5. In the feeding place or in a cocoon on a leaf.

I. 5-6. Flies round the foodplant, especially in the late afternoon.

131(1257). **L. morosa** (Zell.) (*quadripunctella* sensu Steph.)

O. Possibly in flowers of *Rosa* spp.

L. ?-5. In a shoot of rose, causing it to droop.

P. 5. In the feeding place.

I. 5-6. Flies round the foodplant in sunshine from about 0700-1100 hrs; later, rests low down on the branches or on adjacent herbage.

132(1259). **L. fuscatella** (Tengst.) (*tenuicornis* (Staint.))

O. On *Betula* spp., preferring young trees.

L. 10-4. In a round gall in a twig, usually at a node.

P. 5. In the gall. The emergence hole is capped over with red-brown frass towards the end of April.

I. 5-6. Flies in the afternoon.

133(1258). **L. pubicornis** (Haw.)

O. On *Rosa pimpinellifolia*.

L. ?-4. Possibly in the flowers when young. After hibernation in the developing shoots, causing distortion.

P. 5. In the shoot.

I. 5-6. Flies in the afternoon sunshine; rests in the herbage near the foodplant and can be smoked out.

134(1270). **Nematopogon swammerdamella** (Linn.)

O.

L. 8-4. On various herbaceous plants and dead leaves; in a flat case tapered anteriorly, made from pieces of dead leaves. Usually overwintering twice.

P. 4-5. In the case.

I. 5-6. In woodland. Flies at dusk but readily disturbed in daytime.

135(1271). **N. schwarziellus** (Zell.) (*panzerella* sensu auctt.)

O. 6. In a stem or petiole of a herbaceous plant.

L. 7-11. In a flat case tapered anteriorly, constructed from debris and fragments of dead leaves; feeding on dead leaves.

P. ?4-5. In the case; some larvae pupate before winter, others in the spring.
I. 5-6. In woodland. Flies in late afternoon.

136(1272). **N. pilella** ([D. & S.])
O.
L.
P.
I. 5-6. On moors or in open woodland, often amongst *Vaccinium myrtillus*. Flies in sunshine.

137(1273). **N. metaxella** (Hübn.)
O. In the stem or petiole of a herbaceous plant.
L. ?-5. On dead leaves and growing plants, from a flat, oblong case constructed of silk mixed with soil particles, later enlarged with pieces of dead leaf.
P. 5-6. In the case.
I. 6-7. In damp woodland, fens and marshes. Flies towards dusk.

138(1260). **Nemophora fasciella** (Fabr.)
O. In flowers of *Ballota nigra*.
L. 9-4. Initially on seeds of *Ballota nigra*; then in a flat case of silk and plant fragments on lower leaves of foodplant; often in large aggregations.
P. 5-6. In the larval case.
I. 7. Flies in sunshine.

139(1261). **N. minimella** ([D. & S.])
O. In flowers of *Succisa* or *Scabiosa*.
L. 8-4. At first in flowers of foodplant; then in a flat case of silk and plant fragments on leaf-litter.
P. 5-6. In the larval case.
I. 7. Flies in sunshine.

140(1262). **N. cupriacella** (Hübn.)
O. In flowers of *Knautia, Scabiosa* or *Succisa*.
L. 8-4. At first in seeds, then in a flat case on lower and fallen leaves of the foodplant.
P. 5-6. In the larval case.
I. 6-7. Flies in sunshine, rests on flowers of foodplant.

141(1263). **N. metallica** (Poda) (*scabiosella* (Scop.))
O. In flowers of *Knautia* or *Scabiosa*.
L. 8-4. At first in seeds; then in a case formed from a seed husk and finally in a flat case on lower and fallen leaves of the foodplant and probably other leaf-litter.
P. 5-6. In the larval case.
I. 6-7. Flies in sunshine.

142(1264). **N. degeerella** (Linn.)
O.
L. ?-3. Early feeding unknown; later, in a flat case on leaf-litter.
P. 4-5. In the larval case.
I. 5-6. Flies at dusk, sometimes in swarms.

143(1265). **Adela cuprella** ([D. & S.])

O. In catkins of *Salix* spp.
L. 9-3. In a flat case on leaf-litter.
P. 3. In the case.
I. 4. Flies in sunshine, swarming around the tops of tall *Salix* bushes.

144(1266). **A. reaumurella** (Linn.) (*viridella* (Scop.))

O. On herbaceous plants.
L. 9-3. In a flat case made of pieces of dead leaves, on leaf-litter.
P. 4-5. In the case.
I. 5-6. Flies in sunshine; males sometimes fly in swarms around trees and bushes.

145(1267). **A. croesella** (Scop.)

O. Probably in a flower of *Ligustrum vulgare, Hippophae rhamnoides*, etc.
L. 8-3. At first in flowers or seeds of foodplant, then in a flat case on leaf-litter.
P. 4-5. In the case.
I. 6. Flies in sunshine.

146(1268). **A. rufimitrella** (Scop.)

O. In a seed-pod of *Cardamine pratensis*. Probably also of *Alliaria petiolata, Sisymbrium officinale, Barbarea vulgaris* or *Cakile*.
L. 7-10. First instar in a seed-pod; then in a flat case of silk and plant fragments on lower leaves.
P. 11-4. In the case.
I. 5-6. Flies in sunshine; visits flowers of foodplants.

147(1269). **A. fibulella** ([D. & S.])

O. In a flower of *Veronica chamaedrys* or *V. officinalis*.
L. 7-4. At first in the seed capsules; then on lower and fallen leaves, in a flat case of silk and plant fragments.
P. 5. In the case.
I. 6. Flies in the sunshine and rests on flowers of the foodplant.

HELIOZELIDAE

148(825, 826). **Heliozela sericiella** (Haw.) (*stanneella*) (Ford, nec Fisch. v. Rösl.))

O. 5-6. Beneath the bark of a young twig of *Quercus* spp.
L. 6-7. Mines from the twig up a petiole into the base of a leaf, where it excavates a small blotch.
P. 8-4. In a portable case, excised from the blotch and attached to detritus on the ground.
I. 4-5. Flies in sunshine round the lower branches of oak and adjacent undergrowth.

149(827). **H. resplendella** (Staint.)

O. 6-7. In the midrib or a lateral rib of a leaf of *Alnus glutinosa*, preferably a young terminal leaf.

L. 6-9. Mines from the midrib up a lateral rib, thence crossing over in a slender gallery to an adjacent rib in which it mines downwards back to the midrib. Finally it mines a blotch in the blade of the leaf. Leaves which have been mined develop a twist.

P. 9-5. In an oval case cut from the blotch and then attached to detritus.

I. 5-7. Flies in sunshine. It is possible that in some years there is a small second generation.

150(828). **H. hammoniella** (Sorh.) (*betulae* (Staint., 1890))

O. 6. Under the bark of a young twig of *Betula*, preferably in the terminal shoot of a seedling.

L. 7-8. In a mine proceeding up the twig, thence up a petiole and into the leaf where it excavates a small blotch.

P. 8-5. In an oval case cut from the blotch, attached to detritus.

I. 5-6. Flies in afternoon sunshine.

151(829). **Antispila pfeifferella** (Hübn.)

O. 6. In a leaf of *Swida* near the margin. The female makes a row of 2 or 3 "practice" slits at right angles to the margin, which are later visible as brown specks.

L. 7-8. In a gallery leading to a large blotch.

P. 8-5. In an oval case cut from the blotch, measuring 5.5 to 7.0 mm in its longer axis. The case lies on or below the surface of the soil.

I. 5. Flies in afternoon sunshine.

152(830). **A. petryi** (Martini) (*treitschkiella* sensu auctt.)

O. 7-8. In a leaf of *Swida* near the margin, without the "practice" slits.

L. 8-10. In a gallery leading to a large blotch.

P. 10-6. In an oval case cut from the blotch, measuring 4.0 - 5.5 mm in its longer axis. The case lies on the surface of the soil.

I. 6-7.

PSYCHIDAE

153(1247). **Narycia monilifera** (Geoffr.)

O. 7. Laid in cracks of bark. Hatches in 4 to 5 weeks.

L. 8-5. On lichens growing on tree-trunks and fences. In a portable subtriangular case, greyish green to black, dusted with lichen.

P. 5. In the case attached to a trunk or fence.

I. 6-7. Rests on trunks.

154(–). **Solenobia triquetrella** (Hübn.)

O. 3-4. Laid inside the larval case.

L. 4-2. On lichens growing on trees or rocks, decaying plants and dead insects. In a triangular case tapered at both ends, decorated with sand, frass and parts of dead insects. Overwinters full-fed.

P. 2-3. In the case.

I. 3-4. Parthenogenetic; apterous. Protrudes halfway from the case on emergence in early morning.

155(1243). **S. inconspicuella** (Staint.)

O. 5-6. Laid inside the larval case. Hatches in 4 to 5 weeks.

L. 6-3. On lichens growing on tree-trunks, fences and rocks. In a subtriangular case tapered at both ends, nearly black in colour, coated with lichen, wood particles and fine sand. Overwinters full-fed.

P. 3-4. In the case.

I. 4-5. Female apterous.

156(1241). **S. douglasii** (Staint.) (*Douglasi* missp.)

O.

L.

P.

I. One specimen taken about 1850 in Birch Wood, Kent. May be conspecific with *B. conspurcatella*.

157(1244). **S. lichenella** (Linn.)

O. 5. Laid inside the larval case. Hatches in 4 to 5 weeks.

L. 6-3. On lichens, moss and decayed vegetable matter on the ground. In a triangular case tapered at both ends, decorated with lichen, wood particles and frass; colour variable. Hibernates full-fed.

P. 3-5. In the case attached to a tree-trunk, fence or rock.

I. 4-5. Parthenogenetic; apterous.

158(1248). **Diplodoma herminata** (Geoffr.) (*marginepunctella* (Steph.))

O. Laid singly in cracks of bark. Hatches in 4 to 5 weeks.

L. 9-5. On decaying leaves, fungi, dead insects and moss or lichen growing at the base or trees or stumps. In a hard triangular inner case enclosed in a shorter, soft case covered with detritus, fragments of dead insects, etc. Usually living two years.

P. 5-6. In the case attached low down to a trunk or stump.

I. 6-7.

159(1240). **Taleporia tubulosa** (Retz.) (*pseudobombycella* (Hübn.))

O. 6-7. Laid in the case. Hatches in 4 to 5 weeks.

L. 8-4. On lichens, decaying leaves and dead insects on the ground. In a long, triangular silken case, sometimes decorated with dead insects at the front end. Sometimes living two years.

P. 4-5. In the case attached to a trunk, etc. up to two metres from the ground.

I. End 5-6. Female apterous.

160(1242). **Bankesia conspurcatella** (Zell.) (*staintoni* (Wals.))

O. 4-5. Laid in the case. Hatches in 4 to 5 weeks.

L. 5-2. On lichens growing on tree-trunks and fences. In a triangular case, tapered at both ends, covered with sand and lichen. Overwinters full-fed.

P. 2-3. In the case attached to a crack in the bark, rock or fence.

I. 3-4. Female apterous.

161(1246). **Bacotia sepium** (Spey.)

O. 7-8. Laid in the case. Hatches in 4 to 5 weeks.

L. 8-5. On lichens growing on trunks and branches of trees and on posts or fences. In a short, cylindrical case, rounded at the posterior end, covered with lichen; projects at right angles from the bark.

P. 5-6. In the case attached to a trunk, etc.

I. 6-7. Female apterous.

162(1245 partim). **Luffia lapidella** (Goeze)

O. 7-8. Laid in the case.

L. 8-5. On lichen growing on rocks and wood in sunny situations. In a cylindrical case, conical at the posterior end, covered with lichen and sand.

P. 5-6.

I. 6-7. Occurs in the Channel Islands. May be a bisexual race of the following species.

163(1245 partim). **L. ferchaultella** (Steph.)

O. 7-8. Laid in the case. Hatches in 4 to 5 weeks.

L. 8-5. Feeding and case similar to those of *L. lapidella*, but damp, shady situations are preferred. The lichen covering of the case sometimes forms rings of different colours.

P. 5-6. In the case attached to the trunk, fence or rock where the larva fed.

I. 6-7. Parthenogenetic; apterous. Often large numbers occur on a single trunk or rock.

164(218). **Psyche casta** (Pall.)
O. 7-8. Laid in the case. Hatches in 4 to 5 weeks.
L. 8-5. On grasses, lichens and decaying vegetation. In a case covered with longitudinally placed bits of grass stem attached only at the fore end and splayed outwards posteriorly.
P. 6. In the case attached to a tree-trunk, fence or herbage.
I. 6-7. Female apterous.

165(219). **P. crassiorella** (Bruand)
O. 7-8. Laid in the case. Hatches in 4 to 5 weeks.
L. 8-5. On grasses. In a case rather similar to that of *P. casta* but slightly larger and covered with stouter pieces of grass stem.
P. 5-6. In the case.
I. 6-7. Female apterous.

166(221,220). **P. betulina** (Zell.) (*eppingella* (Tutt))
O. 7-8. Laid in the pupal skin. Hatches in 4 to 5 weeks.
L. 8-5. On lichens growing on bushes or tree-trunks; blackthorns on the outside of a thicket seem to be preferred. Larvae in captivity will eat leaves of blackthorn and catkins of birch and sallow. In a cylindrical case, rather pointed at the rear end, and covered with fragments of lichen and bark.
P. 6. In the case attached to a trunk or twig.
I. 6-7. Female apterous. *P. betulina* and *P. eppingella* have been found to be conspecific. *Ent. Gaz. 29:* 219-220 (1978).

167(217). **Epichnopterix plumella** ([D. & S.]) (*pulla* (Esp.))
O. 6-7. Laid in the pupal skin. Hatches in 4 to 5 weeks.
L. 7-4. On grasses. In a case covered with longitudinally placed bits of flat grass attached at both ends and hence converging posteriorly.
P. 4-5. In the case affixed to a grass stem.
I. 5-6. Male flies in sunshine from about 1000-1500 hrs. Female apterous.

168(216). **E. retiella** (Newm.)
O. 6-8. Laid in the pupal skin. Hatches in 4 to 5 weeks.
L. 9-4. On *Puccinellia maritima* and other grasses growing on salt-marshes. In a case covered with longitudinally placed fragments of thin grass stem which project beyond the silken tube at the rear end.
P. 5. In the case affixed low down on a grass stem.
I. 5-6. Female apterous. The male flies in sunshine on still days.

169(214). **Acanthopsyche atra** (Linn.) (*opacella* (Herr.-Schäff.))
O. 6-7. Laid in the pupal skin. Hatches in 4 to 5 weeks.
L. 8-4. On *Salix* and grasses, sometimes living two years. In a case covered with longitudinally placed bits of grass stem or leaves of heather.
P. 4-5. In the case.
I. 5-6. The male flies in the afternoon or evening. Female apterous.

170(213). **Pachythelia villosella** (Ochs.)
O. 6-7. Retained in the abdomen of the female, which remains in the case. Hatches in 4 to 5 weeks.
L. 7-5. On *Calluna, Erica* and grasses, usually living two years. In a large case covered with short, longitudinally placed pieces of twig.
P. 5-6. In the case, affixed to heather or a tree-trunk.
I. 6. The male flies in the evening. Female apterous.

171(–). **Canephora unicolor** (Hufn.)
A species widespread on the Continent, but known in Britain only from a single specimen found misidentified in a series of British-taken *Acanthopsyche atra*. (Barrett, C.G., 1895. *The Lepidoptera of the British Isles* 2:343-344).

172(–). **Thyridopteryx ephemeraeformis** (Haw.) *United States National Museum Bulletin* 244 (1964).
A North American species of which the type specimen was supposedly captured in Yorkshire in the eighteenth century.

173(215). **Sterrhopterix fusca** (Haw. 1809) (*hirsutella* sensu Hübn.)
O. 7-8. Laid in the pupal skin. Hatches in 4 to 5 weeks.
L. 8-5. On grasses, *Calluna, Erica* and the foliage of trees and shrubs, living two years. In a somewhat conical case covered with irregularly placed fragments of heather and other vegetable matter.
P. 5-6. In the case affixed to heather, a tree-trunk or a leaf.
I. 6-7. The male flies at night and comes to light. Female apterous.

TINEIDAE

A number of species with larvae which feed on stored products have a variable life-cycle, the duration depending chiefly on temperature: no dates are given for the early stages of these species. The eggs of all species are laid singly or in small groups on the food plant or food material.

The food materials of many species are greatly varied and only the more usual foods are given.

Unless otherwise stated, the pupa may be found in the larval feeding situation.

174(1191). **Morophaga choragella** ([D. & S.]) (*boleti* (Fabr.))
O.
L. 8-5. On *Piptoporus betulinus* or other bracket fungi; in dead wood.

P. 5-6. In the fungus or wood.
I. 7-8. Comes to light.

175(–). **Euplocamus anthracinalis** (Scop.)
O.
L. 8-4. *On fungus on trees or in rotten wood.
P.
I. 5-8. Recorded as a British species by Stephens (1834), but there is no authentic British specimen.

176(1188). **Dryadaula pactolia** Meyr.
O.
L. Probably in a continuous succession of generations, on the wine-cellar fungus, *Rhacodium cellare*, in cellars.
P.
I. 3-11.

177(1234). **Psychoides verhuella** Bruand (*verhuellella* Staint.)
O. On the fern *Phyllitis scolopendrium*; also recorded on *Asplenium* and *Ceterach*.
L. 8-6. On the sporangia with which it covers a portable case resembling a misplaced sorus.
P. 6. In the case, usually attached to a midrib.
I. 6-7.

178(1235). **P. filicivora** (Meyr.)
O. On *Asplenium* spp., *Dryopteris filix-mas, Polystichum setiferum* or, probably, other ferns.
L. On the lower surface of a frond and on sporangia, the feeding being visible from above as brown blotches, under a cover of sporangia spun into an irregular mass; probably in a continuous series of generations.
P. In a tough white cocoon covered with a mass of sporangia, in the feeding position.
I. 5-9. Flies in early morning sunshine.

179(1220). **Lichenovora nigripunctella** (Haw.)
O.
L. 10-5. In a portable case, on lichens on old fences.
P. 6.
I. 7-8. Has been found in and around buildings.

180(1202). **Meessia richardsoni** (Wals.)
O.
L. 8-10. Sometimes living two years. On lichens on rocks and stones, in a rather flat portable case, covered with fragments of lichen.
P. 11-5. In the case, under stones.
I. 6-7.

181(1203). **Infurcitinea argentimaculella** (Staint.)
O.
L. 4-6. On *Lepraria* or, perhaps, other lichens on shady rocks and walls, constructing long silken tubes covered with lichen fragments.
P. 6-7. In the tube.
I. 7. Flies in sunshine.

182(1218,1219). **I. albicomella** (Herr.-Schäff.) (*confusella* sensu auctt.)
O.
L. *On lichens on rocks.
P.
I. 7-8. In rocky coastal localities.

183(1201). **Ischnoscia borreonella** (Millière)
O.
L.
P.
I. 7-8. In rocky coastal localities.

184(1233). **Celestica angustipennis** (Herr.-Schäff.)
O.
L. *On lichens on *Prunus*, or sometimes on its rotten wood.
P.
I. 5-8.

185(1192). **Myrmecozela ochraceella** (Tengst.)
O.
L. 7-5. Probably living two years. On refuse in a nest of the wood-ants *Formica aquilonia* or *F. lugubris*, in a loosely-spun tube.
P. 5-6. In the feeding position.
I. 6-7. Sits around ant-nests after dark; may be disturbed from nests during day.

186(–). **Ateliotum insularis** (Reb.) (*horrealis* Meyr.) *Proc.S.Lond.ent.nat. Hist.Soc.*, **1947-48**:56.
O.
L. *On dried vegetable matter or stored products.
P.
I. Taken in 1937 in a London warehouse containing imported nuts, dates, almonds, etc.; a native of the Canary Islands.

187(–). **Setomorpha rutella** Zell. *Ent.* **78**:88 (1945).
O.
L. On a variety of stored products of vegetable or animal origin; especially on tobacco.
P. In a closely woven inner cocoon enclosed in a loose outer cocoon to which particles of food and excrement adhere.
I. Found in 1944 in a cargo of sunflower-seed meal imported from Argentina.

188(1239). **Lindera tessellatella** Blanchard
O.
L. On floor-sweepings, probably in a continuous succession of generations.
P.
I. 1-12. Found in the British Isles only in a mill at Bootle, Lancashire.

189(1222). **Haplotinea ditella** (P. & M.)
O.
L. On stored grain, rice or ground-nuts.
P.
I. 6-10.

190(1221). **H. insectella** (Fabr.)
O.
L. On stored products of plant or animal origin, fungi, rotten wood, or on poultry-house or stable refuse.
P.
I. 6-8.

191(1229). **Cephimallota angusticostella** (Zell.) (*simplicella* (Zell.))
O.
L.
P.
I. 6-8.

191A(–). **Cephitinea colongella** (Zagulyaev)
The type specimen of this oriental species is labelled "Anglia"; there is no other evidence for its ever having occurred in Britain.

192(1212). **Nemapogon granella** (Linn.) *Ent.* **76**:95-96(1943).
O.
L. On grain and a great variety of other stored vegetable products; also on bracket-fungi (*Piptoporus* and *Trametes*).
P.
I. 3-9.

193(1214). **N. cloacella** (Haw.) *Ent.* **76**:95-96 (1943).
O.
L. On bracket-fungi, especially *Piptoporus*, or on callus-tissue around tree-wounds; rarely on stored products of vegetable origin.
P.
I. 5-9. Probably bivoltine in the south.

194(1216). **N. wolffiella** Karsholt & Nielsen (*albipunctella* Haw.) *Ent. scand.* **7**:151-2(1976).
O.
L. 8-5. In dead wood or fungus.
P.
I. 6-9.

195(1213). **N. personella** (P. & M.) (*infimella* sensu Ford) *Ent.* **76**:95-96(1943).
O.
L. In dead wood or bracket fungi, especially *Trametes* and *Grifola*; occasionally on stored products of vegetable origin.
P.
I. 3-8.

196(1215). **N. ruricolella** (Staint.) *Ent.* **76**:95-96(1943)
O.
L. 9-4. In dead wood or fungus.
P. 5-6.
I. 6-8.

197(1207). **N. clematella** (Fabr.) (*arcella* sensu auctt.)
O.
L. 9-4. In dead wood, on fungus under dead bark or on bracket fungi.

P. 5-6.
I. 6-8.

198(1211). **N. picarella** (Clerck)
O.
L. 9-4. On bracket fungi, especially *Piptoporus betulinus*.
P. 5-6.
I. 6-7.

199(1210). **Archinemapogon laterella** (Thunb.) (*arcuatella* Staint.)
O.
L. 9-5. On the fungus *Fomes fomentarius*, normally in association with the tenebrionid beetle *Bolitophagus reticulatus*; *on other fungi.
P. 5-6.
I. 6-7.

200(1208). **Nemaxera betulinella** (Fabr.) (*corticella* (Curt.))
O.
L. 9-5. In dead wood or fungus.
P. 5-7.
I. 5-8. Two generations a year have been recorded.

201(1209). **Triaxomera parasitella** (Hübn.)
O.
L. 9-4. In dead wood or fungus.
P. 5-6.
I. 5-7.

202(1206). **T. fulvimitrella** (Sodof.)
O.
L. 9-4. In dead wood or fungus, especially *Piptoporus betulinus;* occasionally on callus-tissue around tree-wounds.
P. 5-6.
I. 5-7.

203(1217). **Triaxomasia caprimulgella** (Staint.)
O.
L. 9-5. In dead wood of beech, oak or elm.
P. 5-6.
I. 6-7. Frequents dark corners of sheds in which it may be found hanging in spiders' webs.

204(1194). **Monopis rusticella** (Hübn.)
O.
L. In birds' nests, on dead animals, owl pellets or on a great variety of refuse of animal origin.
P.
I. 5-9. Probably bivoltine in the south.

205(1195). **M. weaverella** (Scott)
O.
L. *In birds' nests or dove-cotes.
P.
I. 5-8.

206(1197). **M. ferruginella** (Hübn.)
O.
L. In birds' nests, on woollen materials in the open or on refuse of plant or animal origin.
P.
I. 5-10.

207(1198). **M. crocicapitella** (Clemens)
O.
L. In birds' nests, on stored products of vegetable origin including flour, corn, felt and textiles, or on a variety of refuse of plant or animal origin.
P.
I. 6-10. Probably sometimes bivoltine.

208(1196). **M. imella** (Hübn.)
O.
L. In birds' nests, on dead animals and woollen materials in the open or on refuse of animal origin.
P.
I. 6-9. Most often in maritime localities; perhaps sometimes bivoltine.

209(1200). **M. monachella** (Hübn.)
O.
L. In birds' nests and on skins, owl pellets, dead animals and other animal refuse.
P.
I. 5-9.

210(1199). **M. fenestratella** (Heyd.)
O.
L. *In dead wood and fungus, on plant refuse and in hornets' nests.
P.
I. 5-8. Not observed in recent years.

211(1193). **Trichophaga tapetzella** (Linn.) (*tapetiella* missp.)
O.
L. In nests of birds and Hymenoptera or on pellets of birds of prey; on furs, horse-hair or textiles of animal fibres, usually in unheated buildings. Feeding from a silken tube amongst the food material.
P.
I. 5-8.

211A(–). **T. mormopis** Meyr. *Ent. Gaz.* **29**:143-144(1978).
O.
L. On feathers; probably also on other stored products or refuse of animal origin.
P.
I. A far-eastern species, a specimen of which was accidentally introduced into Britain in a consignment of feathers.

211B(–). **Ceratophaga orientalis** (Staint.) *Ent. Gaz.* **29**:139-141(1978).
O.
L. On the hooves and horns of dead animals.
P.
I. A far-eastern species accidentally introduced into Britain in 1878 and again in about 1946.

211C(–). **C. haidarabadi** Zagulajev *Ent. Gaz.* **29**:141-142(1978).
O.
L. On the hooves and horns of dead animals.
P.
I. An Indian species accidentally introduced into Britain in about 1946.

212(1204). **Tineola bisselliella** (Hum.) (*bisellliella* missp.)
O.
L. Most commonly on woollen fabrics in houses, and on fur, hair and various materials of animal origin; in the open in nests of birds or Hymenoptera. Feeding from a silken tube.
P. In a spindle-shaped cocoon.
I. 2-9. Common in houses.

213(1224). **Niditinea fuscipunctella** (Haw.)
O.
L. In nests of birds, mammals or Hymenoptera: often abundant in poultry-houses. Indoors on refuse and on stored products of plant or animal origin.
P.
I. 5-9. Probably sometimes bivoltine.

214(1232). **N. piercella** (Bentinck)
O.
L. In birds' nests in holes; *also in nests of moles and Hymenoptera.
P.
I. 6-8. On wooden walls of outbuildings or at light.

Tinea. With the exception of *T. pallescentella*, as far as it known the larvae of all species construct portable silken cases from which they feed. The cases are flattened dorso-ventrally (those of *T. semifulvella* and *T. trinotella* less flattened) and are open at both ends. Pupation takes place in the case.

215(1227A). **Tinea columbariella** Wocke *Ent.* **83**:169(1950).
O.
L. In a portable case in birds' nests, poultry houses, on owl pellets or occasionally on woollen materials indoors.
P.
I. 6-9. Perhaps sometimes bivoltine.

215A(–). **T. fictrix** Meyr. *Ent. Gaz.* **29**:142-143(1978).
O.
L. *On the excrement of bats.
P.
I. A far-eastern species, a single specimen of which was accidentally introduced into Britain in 1974 in a cargo of Australian grain.

216(1227). **T. pellionella** (Linn.)
O.
L. In a portable case, most commonly on woollens, fur, hair or feathers indoors, or in the open in birds' nests or on owl pellets.
P.
I. 6-10. Common in houses.

217(1226). **T. lanella** P. & M.
O.
L. Probably as *T. pellionella*.
P.
I. 6-7. In wool warehouses.

218(1225). **T. translucens** Meyr. (*metonella* P. & M.)
O.
L. In a portable case on skins or wool; frequently on hide-covered drums imported from Africa.
P.
I. 1-12. Continuous brooded, but requires a temperature of 17°C or over to survive.

218A(–). **T. murariella** Staud.
O.
L. On fur, wool, hides, leather, feathers, hooves, pellets of birds of prey, dried insects, etc.
P.
I. 1-12. Continuous brooded in warehouses. A South American species accidentally introduced into Britain. There is no previous reference in British entomological literature.

219(–). **T. dubiella** Staint. (*turicensis* Müller - Rutz)
O.
L. In a portable case on wool, skins or feathers, or in birds' nests; *on owl pellets.
P.
I. 5-9. Common in houses, where it is apparently bivoltine; out of doors it is univoltine, occurring mainly in July.

220(1223). **T. flavescentella** Haw.
O.
L. In a portable case on wool, fur, feathers or dried insects; * in birds' nests.
P.
I. 2-11.

221(1228). **T. pallescentella** Staint.
O.
L. Without a case, on wool, fur, feathers or refuse of animal origin, usually in unheated buildings or in the open; occasionally in wasps' nests.
P. In a tough cocoon.
I. 1-12.

222(1231). **T. semifulvella** (Haw.)
O.
L. 8-5. In a portable case in birds' nests; occasionally on dead animals or on woollen material in the open.
P.
I. 5-9.

223(1230). **T. trinotella** Thunb. (*lapella* Hübn.)
O.
L. 7-5. In a portable case in birds' nests; occasionally on woollen material in the open.
P.
I. 5-8. In two generations in the south.

224(1205). **Tenaga pomiliella** Clemens *Ent.* **62**:34,73(1929); **64**:19 (1931)
O.
L.
P.
I. 6-9.

OCHSENHEIMERIIDAE

225(1236). **Ochsenheimeria mediopectinellus** (Haw.) (*birdella* (Curt.))
O. On *Dactylis, Bromus, Poa* or other grasses.
L. 5. In the stem. When full-fed, it leaves the stem and may be obtained by sweeping.
P. 6. In a white silken cocoon (partly visible) within the sheath of the foodplant.
I. 7-8. Flies from noon to 1400 hrs.

226(1237). **O. bisontella** (L. & Z.)
O. *On *Agropyron, Bromus* or other grasses.
L.
P.
I. 7. Flies freely in morning sunshine; later rests low down on grass stems.

227(1238). **O. vacculella** Fisch.v. Rösl.
O. *8-4. On *Lolium*; probably also on other grasses.
L. 5. In the stem.
P. 6. *In a silken cocoon beneath a leaf of the host-plant.
I. 7-8. Rests under loose bark or in crevices; sometimes in outhouses or houses.

LYONETIIDAE

228(1160). **Leucoptera laburnella** (Staint.)
O. 5;7;8. On the underside of a leaf of *Laburnum* or *Genista tinctoria*.
L. 6;7;9. In a gallery, in which the frass is green, leading to a large blotch with the frass, now black, deposited in a spiral.

P. 6;8;10-4. In a white cocoon on a leaf or amongst detritus.
I. 5;7;8. Only two generations in cool summers, but more or less continuous-brooded in hot seasons. Flies at dusk.

229(1163). **L. wailesella** (Staint.)
O. 5;7-8. On the underside of a leaf of *Genista tinctoria*.
L. 6;8. In a gallery leading to a blotch; the frass is black throughout the mine.
P. 7;9-4. In a white cocoon on a leaf or twig of the foodplant or amongst detritus.
I. 5;7. Can be disturbed from the foodplant by day.

230(1162). **L. spartifoliella** (Hübn.)
O. 7-8. On a twig of *Sarothamnus* or *Genista tinctoria*; the egg is not filled with black frass and drops off soon after the larva has hatched.
L. 9-5. Mining the bark, starting upwards; the early mine is browner and nearer the surface than that of *Trifurcula immundella*.
P. 5-6. In a white cocoon on the stem of the foodplant.
I. 6-7. Flies at dusk.

231(1164). **L. orobi** (Staint.)
O. 5;7. On the underside of a leaf of *Lathyrus montanus*.
L. 6;8. In a blotch with the frass at first concentrated in the centre, but later deposited haphazard.
P. 6-7;8-5. In a white cocoon on a leaf or detritus.
I. 5;7.

232(1165). **L. lathyrifoliella** (Staint.)
O. 5;7-8. On the upperside of a leaf of *Lathyrus sylvestris*.
L. 6;8. In a blotch on the upper surface of the leaf; a small area round the ovum is stained brown.
P. 6-7;8-4. In a white cocoon on a leaf or detritus.
I. 5;7.

233(1166). **L. lotella** (Staint.)
O. 5;6-7. On a leaf of *Lotus corniculatus* or *L. uliginosus*.
L. 5-6;7-8. In a blotch with the frass forming spiral darker markings; can change leaves.
P. 6;9-4. In a white cocoon on a leaf or stem of the foodplant, or on detritus.
I. 5;6-7. Univoltine except in favourable years and situations.

234(1167). **L. scitella** (Zell.)
O. 7-8. On the underside of a leaf of *Crataegus, Malus, Sorbus aucuparia* or some other species of rosaceous tree.
L. 8-9. In a brown blotch with the frass forming darker spiral markings in the centre.
P. 10-5. In a white cocoon on detritus; occasionally on a leaf of the foodplant.
I. 6-7.

235(1161). **Paraleucoptera sinuella** (Reutti) (*susinella* (Herr.-Schäff.))
O. 6;8. On a leaf of *Populus tremula*.
L. 7;9. In a blotch.
P. 7-8;9-5. In a white cocoon.
I. 6;8.

236(1168). **Lyonetia prunifoliella** (Hübn.)

O.	Under the cuticle of a leaf of *Prunus spinosa, P. japonica, Betula* spp. or *Cotoneaster* spp.
L. 7-8.	In a gallery leading to a blotch.
P. 8-9.	In a cocoon suspended by silken threads under a leaf.
I. 10-4.	Not observed in recent years.

237(1169). **L. clerkella** (Linn.)

O. 4-6;6-7; 8-9.	Under the cuticle of a leaf of a rosaceous tree or bush, or *Betula* spp.; the absence of a visible eggshell at the start of the mine helps to distinguish it from that of a nepticulid.
L. 5;7;9-10.	In a long, narrow, sinuous gallery; the larva may be distinguished from a nepticulid by the deeply incised intersegmental divisions.
P. 5-6;7-8; 10.	In a slight cocoon suspended by silken threads on either surface of a leaf.
I. 6;8;10-4.	Overwinters in thatch or evergreens such as yew.

238(1170). **Bedellia somnulentella** (Zell.)

O. 6;8-9.	On either side of a leaf of *Convolvulus arvensis* or *Ipomaea purpurea*.
L. 7-8;9.	In a gallery containing frass leading to a clear blotch; frequently changes leaves. Silk threads below the mined area catch some of the ejected frass.
P. 8;9-10.	Suspended naked from a few silken threads spun on a leaf of the foodplant.
I. 8;10-5.	

Bucculatrix. The larvae at first mine a leaf of their foodplant, this phase with some species lasting until the penultimate instar. They then leave their mines and spin small, circular "moulting cocoons" on the surface of leaves. After ecdysis, they feed externally, generally from the underside, eating small patches of parenchyma but leaving the upper cuticle intact.

239(1176). **Bucculatrix cristatella** Zell.

O.	On the upperside of a leaf of *Achillea millefolium*, close to the margin.
L. 4-5;7.	When young, mines the leaf; later, feeds externally.
P. 5-6;7-8.	In a white, ribbed cocoon on the foodplant or adjacent herbage.
I. 6;8.	

240(1177). **B. nigricomella** Zell.

O.	On the upper surface of a leaf of *Leucanthemum vulgare*.
L. 3-4;7.	When young, mining the leaf in a long gallery, sometimes changing to a fresh mine, often in a second leaf; later, feeding externally.
P. 4-5;7-8.	In a white, ribbed cocoon on a leaf or adjacent herbage.
I. 5-6;8.	Flies in the evening.

241(1178). **B. maritima** Staint.

O.	On the under surface of a leaf of *Aster tripolium*.
L. 4-5;7-8.	When young, mining in a long narrow gallery. In the final instar it may feed as described for the genus, but often continues to mine in a succession of short galleries. Larva with grey-green dorsal line.

P. 5-6;7-8. In a white, ribbed cocoon on a leaf or, more frequently, on a grass stem.
I. 6;8.

242 (–). **B. capreella** Krogerus *Ent. Gaz.* **18**:155-158 (1967).
O. 5-6. On a leaf of *Achillea millefolium*.
L. 6-7. At first mining a leaflet; later grazing externally on the upperside of the tips of leaflets, eating down to the lower epidermis which withers conspicuously.
P. 7. In a white, ribbed cocoon on the foodplant or adjacent herbage.
I. End 7-5.

243(1179). **B. artemisiella** (Herr.-Schäff.) (*artemisiae* Herr.-Schäff.)
O.
L. 4-5;7. On *Artemisia campestris*.
P. 5;7-8.
I. 5-6;8. One specimen taken in 1865.

244(1180). **B. frangulella** (Goeze)
O. 7-8. On the under surface of a leaf of *Frangula* or *Rhamnus*.
L. 8-9. When young, mining the leaf in a tightly-wound spiral, staining the mined area purple; later, externally from the underside.
P. 10-6. In a pale straw-brown, ribbed cocoon on a leaf or detritus.
I. 6-7. Flies in the evening.

245(1181). **B. albedinella** Zell (*boyerella* (Dup.))
O. 6-7. On the under surface of a leaf of *Ulmus* spp., generally close to the midrib.
L. 7-9. When young, in a slender gallery with two to four diverticula; later it feeds externally from the underside.
P. 10-5. In a white cocoon on bark or detritus, lacking, or almost lacking, the ribs characteristic of the genus.
I. 6. Rests on tree-trunks or fences.

246(1182). **B. cidarella** Zell.
O. 7. On the under surface of a leaf of *Alnus glutinosa*, close to a vein.
L. 8-9. When young, in a long, slender gallery in the leaf; later, externally from the underside.
P. 10-4. In a pale yellow-brown, ribbed cocoon on a leaf-stalk or detritus.
I. 5-6. Rests on trunks. In favourable years early larvae may produce adults in August which give rise to a small second generation.

247(1183). **B. thoracella** (Thunb.)
O. 6-7. On the under surface of a leaf of *Tilia cordata*, generally in an angle between veins.
L. 7-8. When young, mining the leaf in a narrow gallery which usually follows a vein for most of its course; later, feeding externally from either side of the leaf.
P. 8-5. In a pinkish brown cocoon on detritus.
I. 6. Comes to light.

248(1184). **B. ulmella** Zell.

O. 6-7;8-9. On the upper surface of leaves of *Quercus* spp., usually close to the midrib.

L. 7;9-10. When young, mining the leaf in a short, contorted gallery; later, feeding externally from the underside.

P. 7-8;11-4. In a whitish, ribbed cocoon on a leaf (first generation only) or a trunk, not always of the host tree.

I. 5-6;8. Rests on trunks or fences.

249(1185). **B. crataegi** Zell.

O. 6-7. On the upper surface of a leaf of *Crataegus* or *Pyrus*, generally beside a vein.

L. 7-8. When young, mining a leaf in a short contorted gallery; later, feeding externally, generally from the upperside.

P. 9-4. In a white, ribbed cocoon on detritus.

I. 5-6. Flies freely in the evening. In favourable years, early larvae may produce adults in September.

250(1186). **B. demaryella** (Dup.)

O. 7. On the under surface of leaves of *Betula* or, less often, *Castanea*; in the west of England and Ireland, also on *Corylus*.

L. 8. When young, mining the leaf in a short gallery which is contorted except where it follows a vein. In its last two instars it feeds externally, generally from the upperside, making two "moulting cocoons".

P. 9-4. In a white, ribbed cocoon on detritus.

I. 5-6.

HIEROXESTIDAE

251(1187). **Oinophila v-flava** (Haw.)

O.

L. 5-6. On fungus growing on walls in cellars, in silken galleries; also on corks. Perhaps in mild districts also on other materials in the open.

P. 6-7.

I. 7-9. Flies in the open in the Isles of Scilly.

252(1189). **Opogona sacchari** (Bojer) (*subcervinella* (Walk.)) *Bull. ent. Res.* **19**:147-166(1928).

O.

L. Usually on banana (*Musa* spp.), feeding on stems, inflorescences and fruits, externally in a silken gallery, or internally.

P. In a silken cocoon covered with food material in the feeding situation.

I 1-12. Imported as a larva with bananas.

253(1190). **O. antistacta** Meyr. *Exot. Micro.* **5**:87(1937).

O.

L. On banana rind, in a slight web.

P.

I. Has been bred from imported bananas.

GRACILLARIIDAE

GRACILLARIINAE

The larvae, when young, mine leaves of their foodplant, first in the epidermis, then, after ecdysis and a change in the structure of the mouth-parts, in the parenchyma. Nos. 268,272 and 284-286 mine throughout the larval period, but most species have a third phase in which they leave their mines and roll or fold leaves or construct cones, within which they feed on the leaf-surface. Two or three such spinnings may be made by each larva. The pupation site is varied, but most species of *Caloptilia* pupate under a glossy silken membrane spun on the surface of a leaf, and most *Parornix* in the folded edge of a fallen leaf. The adults of some species may be found resting by day on tree-trunks or fences; most fly in the evening sunshine; few, other than *Parornix*, come to light.

254(1109).**Caloptilia cuculipennella** (Hübn.)

O. 6. On the upper surface of a leaf of *Ligustrum vulgare*.

L. 7-9. When young, mining the upper surface; in the second phase, the leaf folds upwards, almost concealing the mine. Later, in a neatly constructed cone at the tip of a leaf.

P. 8-9. In a cocoon in the cone. A circular exit-hole is chewed in the leaf and covered with silk.

I. 9-5.

255(1111). **C. populetorum** (Zell.)

O. 6. On either surface of leaves of *Betula* spp., generally on man-high saplings.

L. 7-8. When young, in a long gallery, on either surface, often running from the top to the base of the leaf, where, in the second phase, a blotch is formed. Later in a leaf rolled longitudinally, upperside inwards, in an untidy manner.

P. 7-9. Under a pale green membrane, generally on the underside of a leaf.

I. 8-5. A small second brood may occur in hot summers.

256(1113). **C. elongella** (Linn.)

O. 4-5;6-7. On the upper surface of a leaf of *Alnus glutinosa*, generally on a vein.

L. 5;7. When young, in an irregular epidermal mine on the upperside, leading to a puckered blotch, generally over a vein; later, rolling a leaf downwards longitudinally, making two such spinnings.

P. 5-6;8. Under a silken membrane, usually on the underside of a leaf.

I. 6;9-4. Overwinters in evergreen trees such as yew.

257(1113A). **C. betulicola** (Hering) *Proc.S.Lond. ent.nat. Hist. Soc.* **1946-47**: 163

O. 4-5;6-7. On either surface of a leaf of *Betula* spp.

L. 5;7. In a short gallery in the epidermis leading to a full-depth blotch between veins, contorting the leaf. Later, in a leaf rolled transversely from the tip, not eating the upper epidermis.

P. 5-6;8. Under a shining, greenish white membrane; generally on the underside of a leaf.

I. 6;9-4.

258(–). **C. rufipennella** (Hübn.) *Ent.Rec.* **83**:291-295; **84**:286(1971-72)

O. 5. On the under surface of a leaf of *Acer pseudoplatanus*, close to the midrib.

L. 6-7. When young, in a tiny mine in the angle between veins; later in three successive cones, each rolled downwards at the tip of a lobe.

P. 7. Under a pale, greenish membrane on the underside of a leaf.

I. 8-5. Overwinters in evergreen trees such as yew.

259(1114). **C. azaleella** (Brants)

O. 5-6;8-9. On the underside of a leaf of azalea (*Rhododendron simsii* and *R. indicum*).

L. 6;9. When young, in a gallery in the lower epidermis, leading to a blotch contorting the leaf. Later, in a cone constructed by rolling a leaf downwards from the tip.

P. 7;9-5. Under a white membrane on the underside of a leaf.

I. 5;8.

260(1115 partim). **C. alchimiella** (Scop.)

O. 6;8-9. On the underside of a leaf of *Quercus* spp.

L. 7;9. When young, in a gallery in the lower epidermis which leads to a small square or triangular full-depth mine between veins; later, in three successive cones, each neatly rolled downwards at the tip of a lobe.

P. 7-8;10-4. Under a shining, pale green membrane on the underside of a leaf.

I. 5;8. Rests on trunks or fences; sometimes comes to light.

261(1115 partim). **C. robustella** Jäckh *Ent.Gaz.* **24**:305-308(1973).

O. 6;8-9. On *Quercus* spp.

L. 7;9. Life history not yet distinguished from that of the preceding species.

P. 7-8;10-4.

I.5;8. Rests on trunks or fences; sometimes comes to light.

262(1116). **C. stigmatella** (Fabr.)

O. 6. On the under surface of leaves of *Populus* and *Salix* spp., beside a vein. Very occasionally on *Betula* spp. (twice, ed.).

L. 7-9. When young, in a gallery in the lower epidermis leading to a blotch on the margin, resembling the mine of a *Phyllonorycter* species. Later, in at least two successive cones, each constructed by rolling the tip or edge of the leaf downwards.

P. 9. Under a silk membrane on the underside of a leaf.

I. 9-5. Perhaps bivoltine occasionally, adults having been recorded in late June.

263(1118). **C. falconipennella** (Hübn.)

O. 5-6. On the underside of a leaf of *Alnus glutinosa*.

L. 7-8. When young, in an oval blotch without a clearly visible preliminary gallery, on the underside of the leaf, generally near the margin. Later, in a small fold on the margin, resembling that made by a species of *Parornix*; three such folds are constructed.

P. 8-9. In a small fold on the leaf-margin.

I. 9-5.

264(1119). **C. semifascia** (Haw.)

O. 5-6. On the underside of a leaf of *Acer campestre* or, less often, *A. pseudoplatanus*, close to a vein.

L. 6-7. When young, in a gallery in the lower epidermis leading to a small squarish blotch. Later, on maple the larva rolls a leaf transversely from the tip, and seals the ends of the tube so formed with the adjacent lobes; on sycamore the larva makes a neat cone.

P. 8. Under a silk membrane on the underside of a leaf.

I. 8-5. Occasionally there is a small second generation of larvae in late August-September, producing adults in October.

265(1117). **C. hemidactylella** ([D. & S.])

O. On *Acer pseudoplatanus*.

L. 7-8. Mining when young; later in a conically rolled leaf.

P. 8-9.

I. 9-5. Status in Britain uncertain.

266(1110). **C. leucapennella** (Steph.) (*sulphurella* (Haw.) nec (Fabr.))

O. 5-6. On the under surface of leaves of *Quercus*, including *Q. ilex*.

L. 6-9. When young, in a gallery in the lower epidermis, formed more as a furrow on holm oak. Later, in a conically rolled leaf; generally a tender, terminal leaf when on holm oak.

P. 7-9. Under a silk membrane on the underside of a leaf.

I. 9-5. Apparently in a single generation extending over a long period.

267(1108). **C. syringella** (Fabr.)

O. 5-6;7-8. Laid singly or in a row of 2-8 against the midrib of a leaf of *Ligustrum, Fraxinus,* or *Syringa;* occasionally on *Jasminum officinale.*

L. 6;8-9. When young, in a gallery in the lower epidermis, leading to a full-depth blotch usually occupied by more than one larva. Later in two successive rather untidy rolls or cones.

P. 6-7;10-4. Under a silken membrane on the underside of a leaf.

I. 4-5;7.

268(1112). **Aspilapteryx tringipennella** (Zell.)

O. 5-6;9. On the underside of a leaf of *Plantago lanceolata.*

L. 6-7;10-4. When young, in a tortuous gallery in the lower epidermis; later, this is developed into a large blotch on the upper surface, causing the leaf to close over the mine.

P. 4-5;7-8. In a cocoon in the mine.

I. 5;8.

269(1120). **Calybites pyrenaeella** (Chrétien). *Ent.* **66**:230(1933); *Ent. Rec.* **74**: 120-121(1962).

O. 5-6. On *Acer campestre*.

L. 6. In a conically rolled leaf, similar to that of *C. semifascia,* but occurring up to a month earlier than that species.

P. 7. Under a silk membrane on the underside of a leaf.

I. 8-5.

270(1107). **C. phasianipennella** (Hübn.)

O. 6. On the underside of a leaf of *Polygonum hydropiper, P. persicaria, Rumex acetosa, R. acetosella* or *Lysimachia vulgaris*.

L. 7-9. When young, in a gallery on the under surface, which is then developed into a blotch on the upperside, resembling that made by a species of *Phyllonorycter*. Later, in a conical roll made from a strip partially cut from the edge of a leaf, each larva making two or more such rolls.

P. 9. In a flimsy silken cocoon in the last roll in which the larva has fed.

I. 9-5.

271(1106). **C. auroguttella** (Steph.)

O. 5-6;8-9. On the under surface of a leaf of *Hypericum perforatum* or *H. pulchrum*.

L. 6;9-10. When young, in a tortuous gallery in the lower epidermis, which is later developed into a blotch which contorts the leaf; lastly, in a leaf spun downwards into a rather twisted cone.

P. 7-8;10-4. In a cocoon in a folded leaf on the ground.

I. 5;8.

[*Micrurapteryx kollariella*(Zell.) was mistakenly placed on the British list, there being no evidence that it has ever occurred in this country.]

272(1096). **Parectopa ononidis** (Zell.)

O. On the under surface of leaves of *Ononis* spp. or *Trifolium* spp., hatching in about 14 days (summer broods).

L. 4-5;6; 7-8. When young, in a gallery in the lower epidermis. Later, the larva eats through to the upper surface and makes a branched blotch astride the midrib. The larva can change leaves.

P. 5-6;6-7; 8. Under a white silken membrane, generally on the upper side of a leaf. Dates are those given by Ford (1949); the autumn larvae probably overwinter and there may be only two generations.

I. 5-6;7;8.

273(1098). **Parornix loganella** (Staint.)

O. On leaves of *Betula* spp.

L. The early stages have not been distinguished from those of *P. betulae*.

P.

I.

274(1105). **P. betulae** (Staint., 1854)

O. 5-6;8-9. On the underside of leaves of *Betula* spp., usually on seedlings.

L. 6;9-10. When young, in a gallery in the lower epidermis leading to a small, *Phyllonorycter*-type mine in which the lower epidermis is discoloured brown. Later, in at least two successive downward folds, each at the tip or on the margin of a leaf.

P. 7-8;10-4. In a folded edge of a leaf, often on the ground.

I. 5;8.

275(1099). **P. fagivora** (Frey.)
O. 6;8-9. On a leaf of *Fagus* or *Carpinus*, generally on the undersurface beside a vein.
L. 7;9. When young, in a gallery in the lower epidermis which is later developed into a small blotch above which the upper epidermis turns brown. Later, the larva folds the tip or edge of a leaf downwards and feeds within, making two or three such folds.
P. 7-8;9-4. In a white cocoon, under the edge of a leaf folded upwards, often a fallen leaf.
I. 5;8. The *Parornix* feeding in hornbeam has been supposed to be a distinct species under the name of *P. carpinella* (Frey).

276(1100). **P. anglicella** (Staint.)
O. 6;8. On the underside of a leaf of *Crataegus, Sorbus torminalis* or, occasionally, *Fragaria vesca*.
L. 7;8-9. When young, in a small, *Phyllonorycter*-type mine on the margin of the leaf, distinguished from the mine of *P. oxyacanthae* by the brown discolouration of the lower epidermis. Later, in at least two successive cones, each spun downwards on a lobe.
P. 7-8;9-4. In a folded edge of a leaf on the ground.
I. 5;8.

277(1101). **P. devoniella** (Staint.) (*avellanella* (Staint.)).
O. 6;8-9. On the under surface of a leaf of *Corylus*.
L. 7;9. When young, in a gallery in the lower epidermis, leading to a squarish blotch causing brown discolouration of the lower epidermis. Later, in at least two successive folds constructed upwards on the leaf-margin.
P. 7-8;9-4. In a folded edge of a leaf on the ground.
I. 5;8.

278(1104). **P. scoticella** (Staint.)
O. 6;8. On the under surface of a leaf of *Sorbus aucuparia, S. aria* or *Malus*, including cultivars.
L. 7;8-9. When young, in a *Phyllonorycter*-like mine on the underside, colouring the epidermis brown. Later, under a dense silken pad in the centre or at the edge of a leaf, which causes the leaf to pucker and resembles a mine.
P. 7-8;9-4. In a folded edge of a leaf, frequently on the ground.
I. 5;8. Bivoltine only in the south.

279(–). **P. alpicola** (Wocke) *Ent. Gaz.* **18**:69-72(1967)
O. On *Dryas*.
L. 7-3. Mining the leaves; in the first leaf, in a flat mine on one side of the midrib; in the second, mining the whole leaf and causing it to fold downwards into a pod; the white lower epidermis of the leaf projects slightly, looking like silken spinning. Overwinters unchanged in its cocoon.
P. 4-5. In a white cocoon on the underside of a leaf.
I. 5.

280(–). **P. leucostola** Pelham-Clinton *Ent. Gaz.* **15**:51-54(1964)
O. On *Dryas*.
L. 7-3. The early stages are not yet differentiated from those of the preceding species.

P. 4-5.
I. 5.

281(1102). **P. finitimella** (Zell.)
O. 5-6;8-9. On *Prunus spinosa*.
L. 6-7;9. The early stages are not yet differentiated from those of the next species, except that the blotch appears to be less contracted by spinning.
P. 7-8;9-4.
I. 5;8.

282(1103). **P. torquillella** (Zell.)
O. 5-6;8-9. On the under surface of a leaf of *Prunus spinosa*, usually close to the midrib.
L. 6-7;9. When young, in a gallery in the lower epidermis, leading to a small, *Phyllonorycter*-type blotch between veins, contorting the leaf. Later within the downwards-folded tip or edge of a leaf, each larva making two or three such folds.
P. 7-8;9-4. In an orange-yellow cocoon, spun on detritus or in the folded edge of a fallen leaf; recorded in birds' nests.
I. 5;8.

283(1097). **Callisto denticulella** (Thunb.) (*guttea* Haw.)
O. 6-7. On the upper surface of a leaf of *Malus*, including cultivars.
L. 7-8. When young, in a fine gallery in the upper epidermis, which is later developed into a blotch resembling that of *Phyllonorycter corylifoliella*, except that it is between veins rather than over one, and readily breaks under tension. Later, in two or three successive folds spun on the edge of a leaf.
P. 9-4. In a folded edge of a leaf on the ground.
I. 5-6.

284(1094). **Acrocercops imperialella** (Zell.)
O. 6-7. On the underside of a leaf of *Symphytum officinale* or *Pulmonaria officinalis*, close to a rib.
L. 7-9. When young, in a gallery in the lower epidermis which is later developed into a blotch; the larva eats the parenchyma only in its last instar, and is virtually invisible until it turns crimson shortly before pupation. Lack of discolouration distinguishes the mines from those of Diptera on the same plant.
P. 8-4. In a brown, silken cocoon on detritus or, occasionally, within the mine.
I. 5-6. In one long, extended generation. Ford (1949) gives *Lathyrus niger* as an additional foodplant; this is the foodplant of *A. hofmanniella* (Schleich), recorded as British through confusion with this species.

285(1095). **A. brongniardella** (Fabr.)
O. 5-6. On the upperside of leaves of *Quercus*, including *Q. ilex*.
L. 6. When young, in a tortuous gallery in the upper epidermis, leading into a large silvery, epidermal blotch; often several galleries converge into the same blotch. The larva eats the parenchyma only in its last instar.
P. 7. In a cocoon spun on detritus.
I. 7-5. Fond of hibernating in thatch.

286(1093). **Leucospilapteryx omissella** (Staint.)

O. 6;8. On the under surface of a leaf of *Artemisia vulgaris*.

L. 7;8-9. When young, in a long epidermal gallery on the underside. Later, in an inflated blotch on the upperside, turning the upper epidermis first yellowish and then purple.

P. 7-8;10-4. In a pale, yellowish brown silken cocoon spun on detritus.

I. 5;8.

LITHOCOLLETINAE

Phyllonorycter. Larvae of all species of this genus feed by mining, first in the epidermis on the sap and later, after ecdysis and a change of mouth-parts, on the parenchyma. The mine is then a small blotch on the upper or lower side of the leaf, made by separating the cuticle from the parenchyma; the separated cuticle is silk-lined and caused to contract, thus forming a pucker or hollow chamber in the leaf. The separated cuticle appears whitish, the mined area on the other side of the leaf being mainly or wholly green. The side mined is constant for each species, apart from very rare aberrant instances (e.g. in nos 313 and 333). Similar mines are made by larvae of *Caloptilia* and *Parornix* in their early instars, but these are less extensively silk-lined, the separated cuticle readily splits if the mine is stretched and, in most species, it turns brown. Nos 311 and 312 differ in that they mine the tender bark of twigs. The method of pupation is constant for each species and gives good diagnostic characters. Some larvae spin cocoons within the mine, in certain instances incorporating frass in the spinning; others construct a pupal chamber, i.e. they thickly line the walls of a section of the mine with silk; a few pupate naked without any spinning. All pupate within the mine, the pupa being extruded on the emergence of the adult. Adults of the tree-feeding species may frequently be found by day resting on the trunks or leaves of the foodplant or neighbouring plants; they are readily disturbed. Most fly in evening sunshine and a few come to light.

287(1042). **Phyllonorycter harrisella** (Linn.) (*cramerella* Fabr.)

O. 5-6;8-9. On *Quercus* spp.

L. 6-7;9-10. In a small mine, usually less than 14 mm long, on the underside of a leaf.

P. 7-8;10-4. In a frass-lined cocoon.

I. 4-5;8-9.

288(1041). **P. roboris** (Zell.)

O. 6-7. On *Quercus* spp.

L. 7-8. In a large oval mine on the underside, contorting the leaf strongly; the lower epidermis is uncreased and there is a pronounced central green patch on the upperside.

P. 9-5. In a cocoon attached solely to the green patch, with the frass stacked behind the cocoon.

I. 6.

289(1044). **P. heegeriella** (Zell.)

O. 6;8-9. On *Quercus* spp.

L. 7;9-10. In a mine on the underside of a leaf, generally at the edge of a lobe causing it to fold right over, and having numerous creases in the lower epidermis. It is the smallest mine on oak, usually less than 10 mm long.

P. 7-8;10-5. In a white cocoon occupying most of the mine; frass stacked behind the cocoon.

I. 5;8. First generation usually emerges a week or two later than *P. harrisella*.

290(1043). **P. tenerella** (Joannis) (*tenella* (Zell.) nec (Dup.))

O. 5-6;8-9. On *Carpinus*.

L. 6-7;9-10. In a mine on the underside of a leaf, usually extending from the midrib to the margin between veins, almost forming a tube.

P. 7-8;10-5. In a white cocoon.

I. 5;8.

291(1088). **P. saportella** (Dup.) (*hortella* Fabr.)

O. 6;8-9. On *Quercus* spp.

L. 7;9-10. In a mine on the underside of a leaf, possibly on the upper part of the tree.

P. 7-8;10-5.

I. 5;8. Has been found on trunks of oak. Not recorded in recent years.

292(1046). **P. quercifoliella** (Zell.)

O. 6;9. On *Quercus* spp.

L. 7;9-10. In a mine on the underside of a leaf.

P. 7-8;10-4. In a white frass-lined cocoon, the frass forming a distinct U-shape when viewed against the light.

I. 4-5;8-9.

293(1045). **P. messaniella** (Zell.)

O. 6;9;11. On *Quercus*, especially *Q. ilex*; occasionally on *Castanea, Carpinus, Fagus* or *Tilia*.

L. 3-4;7;10. In a mine on the underside of a leaf; when on oak, very similar to that of the preceding species.

P. 4-5;7-8; 10-11. Does not overwinter in the pupal stage.

I. 5;8;11. Comes to light more readily than most other members of the genus.

294(1087). **P. muelleriella** (Zell.) (*amyotella* (Dup.))

O. 6;8-9. On *Quercus* spp.

L. 7;9-10. In a mine on the underside of a leaf, frequently extending between lateral veins from the midrib; often two or three mines occur side by side. In most cases the lower epidermis has a central crease.

P. 7-8;10-5. In a white, frass-lined cocoon, firmly attached to the upper surface of the mine.

I. 5;8.

295(1061). **P. oxyacanthae** (Frey)

O. 6;8-9. On *Crataegus*.

L. 7;9-10. In a mine on the underside of a leaf, often in a lobe causing its sides or the tip to fold downwards.

P. 7-8;10-5. In a light brown cocoon, with the frass stacked at one side.

I. 5;8.

296(1056). **P. sorbi** (Frey)
O. 5;8-9. On *Sorbus aucuparia* or, occasionally, *Prunus padus*.
L. 6-7;9-10. In a large mine on the underside of a leaf, often causing the edge of the leaf to fold over; a strong crease in the lower epidermis.
P. 7-8;10-4.
I. 4-5;8.

297(1058,1057). **P. mespilella** (Hübn.) (*pyrivorella* (Bankes))
O. 6;8-9. On *Sorbus torminalis* or *Pyrus*; occasionally on *Sorbus aucuparia, S. aria* or some other rosaceous tree or bush.
L. 7;9-10. In a mine on the underside of a leaf.
P. 7-8;10-5.
I. 5;8.

298(1059). **P. blancardella** (Fabr.) (*concomitella* (Bankes))
O. 5-6;8-9. On *Malus*, including cultivars.
L. 6-7;9-10. In a mine on the underside of a leaf.
P. 7-8;10-5. In a white silk-lined chamber, without frass.
I. 5;8.

299(1060). **P. cydoniella** ([D. & S.]) (*blancardella* sensu P. & M.)
O. 5-6;8-9. On *Cydonia*; less often on *Malus*.
L. 6-7;9-10. In a mine on the underside of a leaf.
P. 7-8;10-5. In a white, silk-lined chamber with the grass stacked behind.
I. 5;8. The relative rarity of its main foodplant probably accounts for the apparent scarcity of this species in Britain.

300(1064). **P. junoniella** (Zell.)
O. 6-7;9. On *Vaccinium vitis-idaea*, especially on small plants growing amongst heather.
L. 7;10-4. In a mine on the underside of a leaf, often occupying its whole area and causing the upper surface to become convex.
P. 5;7-8.
I. 6-7;8-9.

301(1054). **P. pomonella** (Zell.) (*spinicolella* Zell.)
O. 6;8-9. On *Prunus* spp., especially *P. spinosa*.
L. 7;9-4. In a mine on the underside of a leaf; when in a small leaf, contorting it strongly. The second generation overwinters unchanged in the mine.
P. 4;7-8.
I. 5;8.

302(1055). **P. cerasicolella** (Herr.-Schäff.)
O. 6;8-9. On *Prunus avium* or *P. cerasus*.
L. 7;9-4. In a mine on the underside of a leaf. The second generation overwinters unchanged in the mine.
P. 4;7-8.
I. 5;8.

303(1062). **P. lantanella** (Schrank)
O. 6;8-9. On *Viburnum lantana*; sometimes on *V. opulus* or *V. tinus*; very occasionally on *Sorbus aucuparia*.
L. 7;9-4. In a mine on the underside of a leaf, often several in one leaf causing distortion. The second generation overwinters in the mine.

P. 4;7-8. In a silk-lined chamber from which frass is excluded.
I. 5;8.

304(1085). **P. corylifoliella** (Hübn.) (*betulae* (Zell.))
O. 6-7;8-9. On *Malus, Crataegus, Pyrus, Sorbus* spp. or, less usually, another rosaceous tree; f.*betulae* on *Betula* spp.
L. 7;9-10. In a mine on the upperside of a leaf, usually over a vein, causing it eventually to fold upwards; the upper epidermis is at first silver-white, but is later covered with tiny, rust-coloured flecks. The northern and western f. *betulae* occurs 9-10 only.
P. 7-8;10-5.
I. 5-6;8. f. *betulae* is univoltine and may be a distinct species.

305(1084). **P. viminiella** (Sirc.)
O. 6;8-9. On *Salix viminalis* or *S. fragilis*; less often on other *Salix* spp.
L. 6-7;9-10. In a mine on the underside of a leaf, usually near its base, causing the margin to fold over.
P. 7-8;10-5.
I. 5-6;8. Apparently confined to fenlands and other wet situations.

306(1066). **P. viminetorum** (Staint.)
O. 6-7;8-9. On *Salix viminalis*.
L. 7;9-10. In a mine on the margin of the underside of a leaf, causing it to fold downwards.
P. 7-8;10-5. In the mine without a cocoon.
I. 5-6;8. A very local species, apparently restricted to osier beds.

307(1065). **P. salicicolella** (Sirc.)
O. 6;8-9. On *Salix* spp.
L. 7;9-10. In a large mine extending between two veins on the underside of a leaf.
P. 7-8;10-5.
I. 5;8.

308(–). **P. dubitella** (Herr.-Schäff.)
O. 6;8-9. On *Salix caprea*.
L. 7;9-10. In a mine on the underside of a leaf, closely resembling that of the preceding species.
P. 7-8;10-5.
I. 5-6;8.

309(1068). **P. spinolella** (Dup.)
O. 6;8-9. On *Salix* spp.
L. 7;9-10. In a mine on the underside of a leaf, usually near the margin, but not contorting it strongly.
P. 7-8;10-5.
I. 5-6;8. First generation emerges about a fortnight later than *P. salicicolella*.

310(1067). **P. cavella** (Zell.)
O. 8. On *Betula* spp.
L. 9-10. In rather a large mine on the underside, occupying much of the leaf and strongly contorting it; 7-12 creases in the lower epidermis.

P. 10-6. In a cocoon.
I. 6-7. Univoltine, except possibly in the south-east.

311(1049). **P. ulicicolella** (Staint.)
O. 8. On *Ulex europaeus*.
L. 9-5. In the thinner stems, in a narrow mine just under the surface of the green bark, causing slight blistering.
P. 5-6.
I. 6-7. Can be dislodged by beating gorse bushes during the day.

312(1050). **P. scopariella** (Zell.)
O. 8. On *Sarothamnus*, on the bark of new growth, usually of small bushes.
L. 9-5. In a narrow mine in the green bark, causing slight blistering; generally situated 100-200mm from the tip of a twig.
P. 5-6.
I. 6-7. Can be dislodged by beating the foodplant during the day.

313(1053). **P. maestingella** (Müll.) (*faginella* (Zell.))
O. 6-7;8-9. On *Fagus*.
L. 7;9-10. In a long mine on the underside of a leaf, between two veins, or at the margin of the leaf causing it to fold downwards. Occasionally the mine is on the upperside.
P. 7-8;11-5.
I. 5-6;8.

314(1052). **P. coryli** (Nic.)
O. 6;8-9. On *Corylus*.
L. 7;9-10. In a mine on the upperside, the surface of which has a fine, papery appearance; often there are several mines in one leaf.
P. 7-8;10-5.
I. 5;8.

315(1051). **P. quinnata** (Geoffr.) (*carpinicolella* (Staint.))
O. 6;8-9. On *Carpinus*.
L. 7;9-10. In a mine on the upperside of a leaf, eventually causing it to fold upwards.
P. 7-8;10-5.
I. 5;8.

316(1047). **P. rajella** (Linn.) (*alnifoliella* (Hübn.))
O. 6;8-9. On *Alnus*, including *A. cordata*.
L. 7;9-10. In a mine on the underside of a leaf, usually near the midrib and between two veins; often there are several mines in one leaf.
P. 7-8;10-5. In a cocoon incorporating frass.
I. 5;8.

317(1073). **P. strigulatella** (L. & Z.) (*rajella* sensu auctt.)
O. 6;8-9. On *Alnus incana*.
L. 7;9-10. In a mine on the underside, often several in one leaf.
P. 7-8;10-5.
I. 5;8.

318(1048). **P. distentella** (Zell.)
O. 6;8-9. On *Quercus* spp.

L. 7;9-10. In a very large mine on the underside of a leaf, causing a strong contraction of the lower surface to form a tube.
P. 7-8;10-5. In the mine without a cocoon.
I. 5;8.

319(1070). **P. anderidae** (Fletch.)
O. 6;8-9. On *Betula*, especially on seedling bushes amongst heather or on recently cleared paths.
L. 7;9-10. In a mine on the underside of a leaf.
P. 7-8;10-5. In the mine without a cocoon.
I. 5;8.

320(1063). **P. quinqueguttella** (Staint.)
O. 6;8-9. On *Salix repens* or *S. arenaria*.
L. 7;9-10. In a mine on the underside, occupying nearly the whole leaf and causing the edges to curve downwards.
P. 7-8;10-5.
I. 5;8.

321(1071). **P. nigrescentella** (Log.)
O. 5-6;9. On *Vicia sepium*; reputedly on *Trifolium* spp. and *Medicago* spp.
L. 7;9-10. In a mine on the underside, usually occupying the whole leaflet and causing the upper surface to curve downwards at the edges. In a tenanted plant, usually many leaflets contain mines, especially those well buried by grass.
P. 7-8;10-4.
I. 4-5;8-9.

322(1072). **P. insignitella** (Zell.)
O. 6;8-9. On *Trifolium* spp. or *Ononis repens*.
L. 7;9-10. In a mine on the underside of a leaflet, occupying its whole area.
P. 7-8;10-5.
I. 5;8.

323(1074). **P. lautella** (Zell.)
O. 6;8-9. On *Quercus* spp., especially on seedlings.
L. 7;9-10. In a large, strongly inflated mine, extending from the mid-rib between two veins, on the underside of a leaf; often there are several mines in a leaf.
P. 7-8;10-5.
I. 5;8.

324(1075). **P. schreberella** (Fabr.)
O. 6;8-9. On *Ulmus procera* or *U. carpinifolia*, often on a lower leaf.
L. 7;9-10. In a somewhat circular, strongly inflated mine on the underside of a leaf.
P. 7-8;10-5. In a greenish cocoon.
I. 5;8.

325(1069). **P. ulmifoliella** (Hübn.)
O. 6;8-9. On *Betula* spp.
L. 7;9-10. In a comparatively small mine on the underside of a leaf, occasionally folding its edge strongly; 1-6 creases in the lower epidermis.

P. 7-8;10-5. In a cocoon in the mine.
I. 5;8.

326(1077). **P. emberizaepenella** (Bouch.)
O. 6;8-9. On *Lonicera periclymenum*; sometimes on *Symphoricarpus rivularis*; occasionally on *Leycesteria formosa*.
L. 7;9-10. In a very large mine occupying the greater part of the underside of a leaf.
P. 7-8;10-5. In a hard brown cocoon in the mine.
I. 5;8.

327(1078). **P. scabiosella** (Dougl.)
O. 6;9. On *Scabiosa*, generally in young plants.
L. 7-8;10-4. In a mine on the underside of a leaf, usually a radical leaf buried in grass.
P. 8;4-5.
I. 5;8-9.

328(1079). **P. tristrigella** (Haw.)
O. 6;8-9. On *Ulmus*.
L. 7;9-10. In a long, thin tubular mine between veins on the underside of a leaf.
P. 7-8;10-5. In a brown cocoon.
I. 5;8.

329(1080). **P. stettinensis** (Nic.)
O. 6;9. On *Alnus glutinosa*.
L. 7;10. In a mine on the upperside of a leaf, the mine-surface has a thin, tissue-like appearance and is creased.
P. 7-8;10-5.
I. 5;8.

330(1081). **P. froelichiella** (Zell.)
O. 6;8-9. On *Alnus glutinosa*.
L. 7;9-10. In a mine on the underside of a leaf, frequently extending between veins from the midrib to the margin; the largest mine of the alder-feeding species.
P. 7;10-5. In a flimsy, white cocoon in the mine.
I. 5;7-8.

331(1082). **P. nicellii** (Staint.) (*nicelli* missp.)
O. 6;8-9. On *Corylus*.
L. 7;9-10. In a mine on the underside of a leaf; the upper surface of the mine is strongly arched and flecked with green.
P. 7-8;10-5.
I. 5;8.

332(1083). **P. kleemannella** (Fabr.)
O. 6;8-9. On *Alnus glutinosa*.
L. 7;9-10. In a small, roundish mine, between veins but seldom reaching the midrib, on the underside of a leaf.
P. 7-8;10-5. Within the mine in a stouter cocoon than that of *P. froelichiella*.
I. 5;8.

333(1076). **P. trifasciella** (Haw.)
O. 6;9;?. On *Lonicera periclymenum*.
L. 3-4;7-8; 10. In a mine on the underside of a leaf, sometimes twisting it into a cone. Occasionally the mine is on the upperside.
P. 4-5;7-8;10.
I. 5;8;11. It is not known in which stage the winter months are passed.

334(1089). **P. sylvella** (Haw.)
O. 6;9. On *Acer campestre*.
L. 7;10. In a mine on the underside of a leaf, often in a lobe causing it to turn over.
P. 7-8;10-5.
I. 5;8.

335(–). **P. platanoidella** (Joannis)
O. 6;9. On *Acer platanoides.*
L. 7;10. In a mine on the underside of a leaf, generally between veins and well away from the margin.
P. 7-8;10-5.
I. 5;8.

336(1090). **P. geniculella** (Rag.)
O. 6;9. On *Acer pseudoplatanus*, on the underside of a leaf.
L. 7;10. In a mine on the underside of the leaf, usually between two lateral veins.
P. 7-8;10-5.
I. 5;8.

337(1086). **P. comparella** (Dup.)
O. 6;8. On *Populus canescens*; perhaps also on *P. alba* or *P. nigra*.
L. 7;9-10. In a mine on the underside of a leaf.
P. 7-8;9-10.
I. 8;10-5. Overwinters.

338(–). **P. sagitella** (Bjerk.) *Ent.Rec.* **89**:106-107(1977)
O. 5-6;8. On *Populus tremula*.
L. 6;8-9. In a mine on the underside of a leaf.
P. 7;10-4.
I. 4-5;7-8.

PHYLLOCNISTIDAE

339(1091). **Phyllocnistis saligna** (Zell.)
O. 4-5;7-8. On the underside of leaves of *Salix purpurea* or, possibly, *S. fragilis*. An ovum on the upperside leads to an aborted mine.
L. 6;8-9. Mining first the lower epidermis of the leaf, then proceeding via the petiole into the bark of the twig and finally returning, again via the petiole, into another leaf. Larvae mining the upperside of the leaf fail to reach the twig.
P. 6-7;8-9. In a cocoon under a silken membrane on the margin of the leaf last mined.
I. 7;9-4. The imago of the second generation sometimes overwinters in thatch or hayricks.

340(1092). **P. unipunctella** (Steph.) (*suffusella* (Zell.))
O. 5;7-8. On either side of a leaf of *Populus nigra*, including *var. italica*, but more often on the upperside.
L. 6;8. In a long, serpentine mine in the epidermis of the leaf, usually on the upperside; no frass is visible. A mined leaf has the appearance of a snail having crawled over it.
P. 6-7;8-9. In a cocoon under a silk membrane on the mined surface at the edge of the leaf, the edge being partly folded over the cocoon.
I. 7;9-4. The imago of the second generation sometimes overwinters in thatch or hayricks.

341(–). **P. xenia** Hering *Ent.Rec.* **88**:161-164;306(1976)
O. 5;8. On the upperside of a leaf of *Populus canescens*, usually a tender, terminal leaf of a sapling.
L. 6-7;8-9. In a long, sinuous mine in the epidermis of the upperside of the leaf, with a dark, thin central line of frass.
P. 7;8-9. In a cocoon under a silk membrane at the edge of a leaf, the edge being partly folded over the cocoon.
I. 7-8;9-4.

CHOREUTIDAE

342(839). **Anthophila fabriciana** (Linn.)
O. On *Urtica* or *Parietaria*.
L. 6-7;8-5. Under a slight web on the leaves.
P. 5-6;7-8. In a thick, silken cocoon.
I. 5-6;7-9. Flies in sunshine.

343(842). **Tebenna bjerkandrella** (Thunb.)
O. On *Pulicaria; *Eryngium, Cirsium* and probably other plants.
L. 6-7.
P. 7-8.
I. 8-9. Flies in sunshine and rests on flowers of foodplant.

344(841). **Choreutis sehestediana** (Fabr.) (*punctosa* (Haw.)) *Ent.* **72**:257 (1939)
O. On *Scutellaria galericulata*.
L. 6;7-8. In a web on the leaves.
P. 7-8;?-4.
I. 5;7-8.

345(840). **C. myllerana** (Fabr.)
O. On *Scutellaria galericulata*, hatching in about 14 days.
L. 6;7-8. In a web on the leaves.
P. 7-8;?-5. In a white, thick silken cocoon.
I. 5;7-8.

346(838). **Eutromula pariana** (Clerck)
O. On *Malus* or *Crataegus*.
L. 5-6;8. On the upper side of a leaf, under a web drawing the edges together.

P. 6-7;8-9. In a thick, silken cocoon on a leaf of the foodplant.
I. 7;9-3.

347(837). **E. diana** (Hübn.)
O. On *Betula* spp.
L. 7-8. On the upperside of a leaf, under a web.
P. 8. In a dense, pointed, white cocoon under a web beneath a leaf.
I. 8-9. Flies in sunshine and visits flowers of *Cirsium arvense*.

GLYPHIPTERIGIDAE

348(845). **Glyphipterix simpliciella** (Steph.) (*fischeriella* (Zell.))
O. On *Dactylis* or *Festuca arundinacea*.
L. 7-8. On the seeds.
P. In the stem.
I. 5-6. Flies in sunshine.

349(846). **G. schoenicolella** Boyd
O. On *Schoenus*.
L. 5-7. On the seeds.
P.
I. 5-9.

350(847). **G. minorella** Snellen (*equitella* sensu auctt.)
O. On *Sedum acre*.
L. 5. In a shoot.
P. 5-6.
I. 6-7.

351(848). **G. forsterella** (Fabr.)
O. On *Carex vulpina* or other *Carex* spp.
L. 8-4. On the seeds.
P. 4-5.
I. 5-6. Feeds on the flowers of the foodplant.

352(849). **G. haworthana** (Steph.)
O. On *Eriophorum* spp.
L. 7-4. On the seeds, spinning the seedhead to nearby vegetation before the winter.
P. 4-5. In the seedhead; seedheads which are still visible in spring nearly always contain larvae or pupae.
I. 5.

353(843). **G. fuscoviridella** (Haw.)
O. On *Luzula campestris*.
L. 3-4. In the stem.
P. 4-5.
I. 5-6. Flies in sunshine.

354(844). **G. thrasonella** (Scop.)
O. On *Juncus* spp.
L. 4-5. In the stem.
P. 5-6.
I. 6-7.

DOUGLASIIDAE

355(888). **Tinagma ocnerostomella** (Staint.)
O. On *Echium vulgare*.
L. 8-10. On the flowers, later entering the stem.
P. 11-6. In the stem.
I. 6-7. A few stems gathered during the winter will usually produce a large number of imagines.

356(–). **T. balteolella** Fisch. v. Rösl. *Ent.Gaz.* **26**:291-293(1975)
O. On *Echium vulgare*.
L. On the flowers.
P. On the ground or in the dead flowers.
I. 5-6.

HELIODINIDAE

357(834). **Heliodines roesella** (Linn.)
O. On *Atriplex* or *Chenopodium* spp.
L. 6. Gregariously, in spun leaves.
P. 6-7.
I. 7-5. Not seen in Britain for the last 150 years.

YPONOMEUTIDAE

358(902). **Argyresthia laevigatella** Herr.-Schäff.
O. Laid singly on a shoot of *Larix* 5-10 cm from the tip. Hatches in 21-28 days.
L. 7-4. In a terminal twig or shoot, killing it.
P. 4-5. In the twig, the stage lasting 26-31 days; a circular silk-covered hole on the underside indicates its presence.
I. 5-6.

359(904). **A. illuminatella** Zell.
O. *On *Abies alba*.
L. *?-4. In a shoot; the larva eats through the inner bark causing the tip to break off. The end is then sealed with silk.
P. *In the twig.
I. 5-6. Status in Britain requires confirmation.

360(903). **A. glabratella** (Zell.)
O. On *Picea abies*.
L. 1-4. In a shoot, discolouring the leaves which later fall off. There is a small circular hole, usually on the underside of an affected shoot a few cm from the tip.
P. 4-6. In the shoot.
I. 6.

361(906). A. **praecocella** Zell.
O. On *Juniperus*.
L. 7-9. In the green berries.
P. 9-5. Under moss or in the ground.
I. 5.

362(905). A. **arceuthina** Zell.
O. On *Juniperus*.
L. ?-4 or 5. In a shoot, causing it to die.
P. 5. In the shoot.
I. 5-6.

363(915). A. **abdominalis** Zell.
O. On *Juniperus*.
L. ?-4 or 5. In leaves, mining from the base to the tip and back, then via the twig to the next leaf; an exit hole is left only in the last leaf to be mined.
P. 5-6. In a web amongst needles on the ground; the stage lasts about three weeks.
I. 7.

364(907). A. **dilectella** Zell.
O. On *Juniperus*.
L. ?-5. In a shoot, causing it to dry out and become discoloured.
P. 6. Outside the mine, in a silken web.
I. 7-8.

365(914). A. **aurulentella** Staint.
O. On *Juniperus*.
L. ?-4 or 5. Mining the leaves; entering from the upper surface, it eats out the apical part of the leaf, sometimes leaving frass behind. It then repeats the process in another leaf, without ever entering the stem. Leaves can be found with one or two holes.
P. 5-6. Outside the mine, usually on the ground; the stage lasts about seven weeks.
I. 7.

366(908). A. **ivella** (Haw.) (*andereggiella* (Dup.))
O. On *Corylus* or *Malus*, especially *M. sylvestris*.
L. 5. In a shoot.
P. 6.
I. 7-8.

367(909). A. **brockeella** (Hübn.)
O. On *Betula* spp. or *Alnus glutinosa*.
L. 10-4. In a shoot or catkins.
P. 5.
I. 6-7.

368(910). A. **goedartella** (Linn.)
O. On *Betula* spp. or *Alnus glutinosa*.
L. 11-4. In a shoot or catkins, sometimes congregating under bark in the spring.
P. 5. In a cocoon in or under the bark.
I. 6-8.

369(911). **A. pygmaeella** (Hübn.)
O. On *Salix caprea* or *S. cinerea*.
L. 4-5. In a shoot or catkins.
P. 5-6.
I. 6-8.

370(912). **A. sorbiella** (Treits.)
O. On *Sorbus aucuparia* or *S. aria*.
L. 5-6. In a shoot.
P. 6.
I. 6-7.

371(913). **A. arcella** (Fabr.) (*cornella* sensu auctt.)
O. On *Malus*.
L. 4-5. In a shoot or flowers.
P. 5-6.
I. 6-7.

372(916). **A. retinella** Zell.
O. On *Betula* spp.
L. 5. In a shoot or bud, causing it to wither.
P. 5-6.
I. 6-7.

373(917). **A. glaucinella** Zell.
O. On *Quercus* spp. or *Aesculus*.
L. 3-4. In the bark, exuding red frass.
P. 5-6. In the bark.
I. 6-7. Emerges in the morning and can be found low down on the trunks until about 1100 hrs.

374(918). **A. mendica** (Haw.)
O. On *Prunus spinosa*.
L. 4. In a flowering shoot.
P. 5.
I. 6.

375(920). **A. conjugella** Zell.
O. On *Sorbus aucuparia, S. aria* or *Malus*; hatches in 12-13 days.
L. 8. In the fruits; in apples, boring into the core.
P. 9-5. In a cocoon under loose bark or in litter on the ground.
I. 6-7.

376(919, 921). **A. semifusca** (Haw.) (*spiniella* Zell.)
O. On *Crataegus* or *Sorbus aucuparia*.
L. 5-6. In a shoot.
P. 6-7.
I. 8.

377.(922). **A. pruniella** (Clerck) (*ephippella* (Fabr.))
O. On *Prunus cerasus*.
L. 4-5. In a shoot.
P. 6.
I. 7.

378(923). **A. curvella** (Linn.) (*nitidella* (Fabr.))
O. On *Crataegus*.
L. 5. In a shoot.
P. 6.
I. 7-8.

379(924). **A. albistria** (Haw.)
O. On *Prunus spinosa*.
L. 5. In a shoot.
P. 6.
I. 7.

380(925). **A. semitestacella** (Curt.)
O. On *Fagus*.
L. 5-6. In a shoot.
P. 7.
I. 8-9.

381(944). **Yponomeuta evonymella** (Linn.)
O. On *Prunus padus*; probably on other *Prunus* spp. in the south.
L. 5-6. On the leaves, in a web; larvae gregarious.
P. 6-7. In a cocoon in the web.
I. 7-8.

382(942 partim). **Y. padella** (Linn.)
O. On *Crataegus, Prunus spinosa* or *P. cerasifera*. Most hatch in about 21 days but a proportion hatch in the spring. Laid in batches on the main stem or on spines and covered with a hardened secretion.
L. 8-6. Autumn larvae overwinter until May under the secretion and then feed gregariously in a web on the leaves, together with those which hatch in the spring.
P. 6-7. In a cocoon in the web.
I. 7-8. Distinct biological races on *Crataegus* and *Prunus*; status uncertain.

383(942 partim). **Y. malinellus** Zell.
O. On *Malus*.
L. 8-6. On the leaves, in a web; larvae gregarious.
P. 6-7. In a cocoon in the web.
I. 7-8. Possibly only a race of *Y. padella*.

384(943). **Y. cagnagella** (Hübn.) (*cognatella* (Treits.))
O. On *Euonymus europaeus* or *E. japonica*.
L. 5-6. On the leaves, in a web; larvae gregarious.
P. 6-7. In a dense cocoon in the web or amongst detritus on the ground.
I. 7-8. Possibly only a race of *Y. padella*.

385(941). **Y. rorrella** (Hübn.) (*rorella* (Hübn.))
O. On *Salix alba* or, occasionally, *S. cinerea*.
L. 5-6. On the leaves, in a web; larvae gregarious.
P. 6-7. In a cocoon in the web or on the trunk.
I. 7-8.

386(940). **Y. irrorella** (Hübn.) *Ent.Gaz.***12**:56(1961)
O. On *Euonymus europaeus*.
L. 8-6. On the leaves, in a web; larvae gregarious until the last instar.
P. 6-7. Usually singly in a flimsy cocoon under a leaf or near the ground.
I. 7-8.

387(939). **Y. plumbella** ([D. & S.])
O. On *Euonymus europaeus*.
L. 5-6. When young, in a shoot causing it to droop; later in spun leaves.
P. 6-7.
I. 7-8.

388(938). **Y. vigintipunctata** (Retz.)
O. On *Sedum telephium*.
L. 6-7;9-10. On the leaves, in a slight web; larvae gregarious. Possibly also feeding on other plants.
P. 11-3;7-8. In a cocoon amongst leaf-litter or spun up under a flower-head of the foodplant.
I. 4-5;8.

389(937). **Euhyponomeuta stannella** (Thunb.) *Ent.***62**:1-2(1929)
O. On *Sedum telephium*.
L. 8-6. Burrowing into the stem and roots, the larva under slight web at the base; up to six larvae in a web; when young, in a silken tube extending into the ground. A grey, withered shoot in the centre of the plant indicates the presence of larva.
P. 6. In a silken cocoon amongst detritus on the ground.
I. 6-7.

390(927). **Kessleria fasciapennella** (Staint.)
O.
L.
P.
I. 9-10.

391(928). **K. saxifragae** (Staint.) *Ent.Gaz.***13**:163-164(1962)
O. On *Saxifraga* spp.
L. 5-6. In a web in the heart of *S. aizoides* or *S. oppositifolia;* in blotch-mines in leaves of *S. hirsuta* or *S. spathularis*.
P. 6.
I. 6-7.

392(926). **Zelleria hepariella** Staint.
O. On *Fraxinus*.
L. 6. In a thick silken web, usually on the young leaves at the tip of a branch.
P. 7. In a dense white silken cocoon on a leaf.
I. 8-4. Hibernates in yew, juniper and other thick cover.

393(929). **Pseudoswammerdamia combinella** (Hübn.)
O. Laid singly on a leaf of *Prunus spinosa*, hatching in about 14 days.

L. 7-9. At first mining a lower leaf making a white blotch from which frass is ejected. Later in a thick web, often several larvae in the same web.
P. 10-4. In a dense white silken cocoon amongst detritus on the ground.
I. 5-6.

394(930). **Swammerdamia caesiella** (Hübn. 1796) (*heroldella* Hübn.)
O. On *Betula* spp.
L. 7;9. Under a web, on the upperside of a leaf. The webs are very conspicuous on dewy mornings.
P. 7-8;9-4. In a dense white silken cocoon in detritus on the ground.
I. 5-6;8.

395(933). **S. pyrella** (Vill.)
O. On *Crataegus, Malus* or *Prunus cerasus*.
L. 7;9. Under a web on the upperside of a leaf.
P. 7-8;9-4. In a dense white silken cocoon on a twig at the base of the leaves or amongst detritus on the ground.
I. 5;8.

396(934). **S. compunctella** (Herr.-Schäff.) *Ent.* **66**:141(1933)
O.
L. Thought to be associated with *Betula* spp. and *Sorbus aucuparia*.
P.
I. 6.

397(932). **Paraswammerdamia spiniella** (Hübn.) (*caesiella* (Hübn., 1810-13)) *Ent.Rec.* **88**:218-219(1976)
O. On *Prunus spinosa*.
L. 9-6. At first the larva mines leaves; it then spins a small white hibernaculum in the axil of a leaf. In the spring in a thick web on the leaves, often two or more larvae in the same web.
P. 7. In a dense white silken cocoon on a twig amongst the leaves.
I. 7-8.

398(931). **P. lutarea** (Haw.)
O. On *Crataegus* or *Sorbus aucuparia*.
L. 9-5. On the leaves in a thick web; when young, several larvae in the same web.
P. 6. In a dense white silken cocoon on a twig amongst the leaves.
I. 7.

399(901). **Cedestis gysseleniella** Zell. (*gysselinella* (Dup.))
O. Near the base of a needle of *Pinus sylvestris*.
L. 3-4. At first mining the needle towards the tip; if necessary, then feeding externally in a very loose web.
P. 5. Amongst needles in a very loose web.
I. 6-7.

400(900). **C. subfasciella** (Steph.) (*farinatella* (Dup.))
O. At the tip of needles of *Pinus* spp., on the upperside.
L. 12-4. Mining the needle towards the base, appearing greenish in the mine.

P. 1-6. In a cocoon on the ground.
I. 3-7. Timing depends on seasonal temperature.

401(899 partim). **Ocnerostoma piniariella** Zell.
O. At the tip of a needle of *Pinus sylvestris*.
L. 4-5. Mining the needle towards the base, appearing, in later instars, yellowish in the mine.
P. 5-6. In a slender cocoon spun between three or four needles.
I. 6.

402(899 partim). **O. friesei** Svensson (*piniariella* sensu auctt., partim) *Opusc. ent.* **31**:196-200; pl.4, figs 1-2(1966)
O. At the tip of a needle of *Pinus sylvestris.*
L. 1-4;6-7. Mining the needle towards the base, appearing, in later instars, yellowish in the mine.
P. 2-5;7. In a slender cocoon spun between three or four needles.
I. 4-5;7-8.

403(952). **Roeslerstammia pronubella** ([D. & S.]) *Ent.* 55:11(1922).
O.
L.
P.
I. 5-6. Taken in Sutherland in 1854.

404(953). **R. erxlebella** (Fabr.)
O. 6-7;9. On the upperside of a leaf of *Tilia*, especially *T. cordata*, or *Betula* spp., close to the margin and generally at the tip.
L. 7;9-10. Mining in the first instar; later exposed on a leaf.
P. 11-4;8. In a cocoon under the turned-down edge of a leaf.
I. 5-6;8-9. Rests high up on trees. Can be obtained by kicking the trees in the morning when the moth flies to the ground; comes freely to light.

405(935). **Atemelia torquatella** (Zell.)
O. On *Betula* or *Ulmus*.
L. 8-3. In a blotch in a leaf, hibernating in the mine.
P. 3-6. In a slight cocoon spun amongst litter on the ground.
I. 6-7.

406(936). **Prays fraxinella** (Bjerk.) (*curtisella* (Don.))
O. On the underside of a leaf of *Fraxinus*, close to the midrib.
L. 10-5. When young, mining a leaf; during the winter, in the bark just below a bud; in spring, within a shoot or in a lightly spun shoot which it causes to droop.
P. 5-6.
I. 6-7.

407(945). **Scythropia crataegella** (Linn.) *Ent.Rec.* **88**:218-219(1976)
O. On *Crataegus, Prunus spinosa* or *Cotoneaster* spp.
L. 9-6. In the autumn, the larva first mines a leaf, then feeds externally, mining again to spin a hibernaculum; in the spring, externally in a web; larvae gregarious.
P. 6-7. In the web, suspended by threads.
I. 7.

Ypsolopha. Although the larvae live in a slight web, they may easily be beaten out of the foodplant, and can be recognised by their spindle-shape.

408(1132). **Ypsolopha mucronella** (Scop.) (*caudella* (Linn.))
O. On *Euonymus europaeus*.
L. 6-8. In a slight web on the leaves.
P. 8. In a boat-shaped cocoon affixed to detritus on the ground.
I. 8-4.

409(1134). **Y. nemorella** (Linn.)
O. On *Lonicera* spp.
L. 5-6. In a slight web on the leaves.
P. 6-7. In a long brownish cocoon amongst detritus or on the foodplant.
I. 7-8.

410(1133). **Y. dentella** (Fabr.) (*xylostella* sensu auctt.)
O. On *Lonicera* spp.
L. 5-6. In a web on the leaves.
P. 6-7. In a rather long white cocoon affixed to detritus on the ground.
I. 7-8.

411(1135). **Y. asperella** (Linn.)
O. *On *Malus* or *Crataegus*.
L. *6. On the leaves, in a slight web.
P. *7.
I. 8-4.

412(1136). **Y. scabrella** (Linn.)
O. On *Malus* or *Crataegus*.
L. 5-6. On the leaves, in a slight web.
P. 6. In a boat-shaped cocoon amongst detritus or under loose bark.
I. 7-8.

413(1137). **Y. horridella** (Treits.)
O. On *Prunus spinosa* or *Malus*.
L. 5-6. On the leaves, in a slight web.
P. 6. In a boat-shaped cocoon amongst detritus on the ground.
I. 7-8.

414(1138). **Y. lucella** (Fabr.)
O. On *Quercus* spp.
L. 5-6. On the leaves, in a slight web.
P. 6-7.
I. 7-8.

415(1139). **Y. alpella** ([D. & S.])
O. On *Quercus* spp.
L. 6. On the leaves, in a slight web.
P. 7. In a boat-shaped cocoon.
I. 8.

416(1140). **Y. sylvella** (Linn.)
O. On *Quercus* spp.
L. 6. On the leaves, in a slight web.
P. 7. In a boat-shaped cocoon.
I. 8-9.

417(1141). **Y. parenthesella** (Linn.) (*costella* (Fabr.))
O. On *Quercus* spp., *Carpinus, Betula* spp., *Corylus* or *Myrica*.
L. 5-6. On the leaves, in a slight web.
P. 7. In a brown boat-shaped cocoon, on a leaf, trunk, or amongst detritus on the ground.
I. 8-9.

418(1142). **Y. ustella** (Clerck) (*radiatella* (Don.))
O. On *Quercus* spp.
L. 5-6. On the leaves, in a slight web.
P. 7. In a whitish boat-shaped cocoon, on a leaf, trunk, or amongst detritus on the ground.
I. 8-4. Fond of resting in gorse or yew in the autumn.

419(1143). **Y. sequella** (Clerck)
O. On *Acer campestre* or, sometimes, *A. pseudoplatanus*.
L. 5-6. On the leaves, in a slight web.
P. 7. In a soft white cocoon in moss on the trunks or amongst detritus on the ground.
I. 8-9. Rests high up on the trees and when dislodged flies down and settles low down on the trunk.

420(1144). **Y. vittella** (Linn.)
O. On *Ulmus* or *Fagus*.
L. 5-6. On the leaves or flowers, in a slight web.
P. 6-7. In a soft white cocoon amongst leaves or amongst detritus on the ground.
I. 7-8.

421(1146). **Plutella xylostella** (Linn.) (*maculipennis* (Curt.))
O. On *Brassica* or other Cruciferae. Laid singly or in twos or threes on the underside of a leaf, hatching in a few days.
L. 6-7;8-9. On the leaves, in a slight web.
P. 7-8;10-4. In an open network cocoon on the foodplant or amongst detritus on the ground.
I. 5;8-9. Populations often reinforced by migration.

422(1145). **P. porrectella** (Linn.)
O. On *Hesperis*.
L. 4;6. On the leaves, in a slight web.
P. 4-5;6-7. In an open network cocoon under a leaf or amongst detritus on the ground.
I. 5;7-8.

423(1149). **Rhigognostis senilella** (Zett.) (*dalella* (Staint.))
O. On *Arabis* spp. or *Descurainia*.
L. 6-7. On the leaves, in a slight web.
P. 7-8. In an open network cocoon.
I. 8-4.

424(1147). **R. annulatella** (Curt.)
O. On *Cochlearia officinalis*.
L. 5-6. Amongst spun flowers and leaves.
P. 6-7. In an open network cocoon amongst detritus on the ground.
I. 7-4.

425(1148). **R. incarnatella** (Steud.)
O. Reputed to be on *Alliaria petiolata*.
L. 6-7. On the leaves in a slight web.
P. 8-9.
I. 9.

426(1131). **Eidophasia messingiella** (Fisch. von Rösl.)
O.
L. 4-5. On *Cardamine amara*, between spun leaves; or on *Cardaria* on the underside of leaves, chewing holes and the edges.
P. 5-6. In an open network cocoon under a leaf of the foodplant or amongst detritus on the ground.
I. 6-7. Flies a little before dusk; strongly attracted to light.

427(1130). **Orthotaelia sparganella** (Thunb.)
O. On *Sparganium* spp., *Iris* spp. or *Glyceria maxima*.
L. 5-6. In the leaves and stem.
P. 7. In the stem.
I. 7-8. Imagines are very sluggish and can be found at rest on the leaves and stems. Pupae are more easily obtained.

428(1154). **Digitivalva perlepidella** (Staint.) *Ent.Rec.* **81**:187-189(1969)
O. On *Inula conyza*.
L. 4-5. In the lower leaves, hollowing them completely; moves via the petioles to a fresh leaf.
P. 5-6. In a cocoon in the mine at the base of the leaf.
I. 5-6.

429(1150). **D. pulicariae** (Klimesch) (*granitella* sensu auctt.)
O. On *Pulicaria dysenterica*.
L. 6-7. In the lower leaves, making brown-white inflated blotches.
P. 7-8. In an open network cocoon under a leaf.
I. 8-5. Hibernates in sheltered situations, e.g. juniper bushes.

430(1155). **Acrolepiopsis assectella** (Zell.) *Ent.* 77:76(1944)
O. On *Allium* spp., especially the cultivated varieties; hatches in 5-8 days.
L. End 5-6; 8-10. On or between the leaves; in leeks, fouling them with frass.
P. 6-7;9-10. In an open network cocoon attached to the foodplant or to detritus nearby.
I. 7;10-5. A potential pest.

431(1151). **A. betulella** (Curt.)
O.
L.
P.
I. 8-10.

432(1152). **A. marcidella** (Curt.)
O.
L.
P.
I. 10-5.

433(1153). **Acrolepia pygmeana** (Haw.) (*pygmaeana* missp.)
O. On *Solanum dulcamara*, especially in sheltered situations.
L. 6;9. In the leaves, making large white blotches.
P. 6-7;9-10. In an open network cocoon amongst detritus on the ground.
I. 7;10-4.

EPERMENIIDAE

434(1125). **Phaulernis dentella** (Zell.)
O. 6-7. On *Chaerophyllum temulentum*, *Pimpinella* or *Aegopodium*.
L. 7-8. On the seeds.
P. 9-5.
I. 6.

435(1124,1123). **P. fulviguttella** (Zell.) (*auromaculata* (Frey))
O. On *Heracleum sphondylium* or *Angelica*.
L. 9-10. In spun seeds.
P. 11-6. In a slight cocoon on the ground.
I. 7-8.

436(1122). **Cataplectica farreni** Wals.
O.
L. 8-9. Probably on the seeds of *Pastinaca*.
P.
I. 6-7.

437(1121). **C. profugella** (Staint.)
O. 8-9. On *Daucus*, *Pimpinella*, *Angelica* or *Aegopodium*.
L. 9-10. On the seeds.
P. 11-6. In a slight cocoon on the ground. In captivity often amongst the seeds.
I. 7-8.

438(1126). **Epermenia illigerella** (Hübn.)
O. On *Angelica* or *Aegopodium*.
L. 5-6;8. First generation in spun leaflets; second in umbels.
P. 6;8. In an open network cocoon amongst detritus on the ground.
I. 6-7;8-9.

439(1129). **E. insecurella** (Staint.)
O. On *Thesium*.
L. 4-5;7. When young mining the leaves; later externally on the leaves and also, in the second generation, on the flowers and unripe seeds.
P. 5;7. In an open network cocoon amongst detritus on the ground or attached to the stem below the surface of the soil.
I. 5-6;7-8.

440(1127). **E. chaerophyllella** (Goeze)
O. 4-5;8. On *Heracleum sphondylium*, *Pastinaca*, *Anthriscus* spp. or *Angelica*.
L. 5-6;8-9. In a slight web on the underside of a leaf, making brown blotches on the upperside; often several larvae on one leaf.
P. 6-7;9-10. In an open network cocoon amongst detritus on the ground.
I. 7-8;10-4.

441(1128). **E. aequidentellus** (Hofm.) (*daucellus* (Pey.))

O. On *Daucus*.

L. 5-6;8-9. On the leaves; when young, mining and making blotches; later externally.

P. 6;9. In an open network cocoon amongst detritus.

I. 6-7;9-10.

SCHRECKENSTEINIIDAE

442(836). **Schreckensteinia festaliella** (Hübn.)

O. 6-7;8-9. On *Rubus* spp.

L. 7;9. Beneath the leaves, skeletonizing them.

P. 8;10-4. In a network cocoon on the underside of a leaf.

I. 4-6;8. Flies in the afternoon.

COLEOPHORIDAE

443(835). **Augasma aeratella** (Zell.)

O. On a flower of *Polygonum aviculare*.

L. 7-4. In a conspicuous, reddened, pod-like gall up to 15 mm long, projecting laterally from the stem; overwinters full-fed in the gall.

P. 5. In the gall. The larva makes an exit-hole shown up by frass just before pupating.

I. 6-7. Last seen in 1934. Old records imply Breckland, downland, woodland margins and cornfields on chalk as habitats.

444(954). **Metriotes lutarea** (Haw.) (*modestella* (Dup.))

O. Possibly on *Stellaria holostea*.

L. Published accounts of the larval habits appear to be fictitious: no larvae have been found on *Stellaria* in Britain. If this be the true foodplant, the pabulum could be fallen seeds or even running stems in leaf-litter.

P.

I. 5. Feeds on the nectar of flowers of the supposed foodplant.

445(955). **Goniodoma limoniella** (Staint.)

O. On a flower of *Limonium vulgare*.

L. 9-5. On the seeds in a portable case made of a floret. When full-fed, makes a short tunnel in the pith of the stem of its host plant or some other thin-stemmed plant where it overwinters. The case at first remains fixed to the stem but may be washed off in winter revealing the entrance hole sealed with silk.

P. 6. In the stem.

I. 7-8. In salt-marshes.

Coleophora. The larvae construct portable cases of silk or of one or more assembled pieces of hollowed-out foodplant; the case may be formed early in larval life or not until the final instar. The larval and (except for nos 545 and 546) pupal stages are passed in the case, which has sprung valvular flaps at its extremity for the ejection of frass and, later, for the eclosion of the imago. The larva fixes the case to a leaf or seed-vessel and bores through the cuticle or husk to feed on the substance within.

A species making its case of a mined leaf hatches from its ribbed, hemispherical egg into the leaf-tissue and makes a gallery a millimetre or two long containing frass. This is expanded into a tiny elongate blotch which is then cut out and sewn round the edge with silk to form the first case. The case is then attached to a leaf, usually on the underside, the larva eating all the cell-layers between the leaf-cuticles as far as it can reach from the case. The result is a series of pale spots without frass, each with a round hole in the lower cuticle of the leaf. These blotches may be closely grouped, or the larva may wander from branch to branch of trees leaving only scattered single mines, particularly at times of ecdysis. Fresh mines are palest. As the larva grows, the case may be enlarged or a new case constructed from a freshly mined section of leaf. Some of the species which make pistol-shaped cases of silk feed as described, but others graze the upper surface of leaves without mining. The angle of the plane of the mouth-opening (hereafter abbreviated m.o.) to the axis of the case is often of diagnostic importance. The case may lie prostrate (m.o.0°), or be vertical (m.o.90°), or be fixed at some characteristic intermediate angle (e.g. m.o.60°).

Species feeding in seedheads have reticulated, plastic eggs which are thrust into corners. On hatching, the larva bores into the seedhead at once and feeds on the unripe seeds. At this phase there is no external sign of attack. When it is necessary to move to another seedhead, a case is constructed, either of silk alone or of the seedhead which has just been eaten out. The larva then bores into a second seedhead, leaving the case attached to the outside. When it has again moved on, the neat, round hole, often with a little ridge of silk from the attachment, is a tell-tale.

Pupation occurs two to four weeks before emergence; most species which are full-fed in the autumn do not choose their pupation site until spring. Miners on trees and bushes tend to pupate on the upper surface of a leaf; those on herbaceous plants near the base of any available stem; seed-feeders on or near the ground.

The imagines of most species are active in late afternoon or evening, but metallic species fly in full sunshine; all groups are also attracted to light at night.

The span of life of coleophorid larvae and their size at hibernation are dependent on temperature and food supply. Life-cycles are poorly synchronized in Britain: a small proportion of individuals of diverse species has been observed to spend two years as larvae and a biennial cycle is normal in a few species. Only *C. frischella* is likely to prove bivoltine in England, though *C. trifolii* is so in southern Europe.

Rearing. Coleophorid larvae will usually feed indoors in a closed container, but their cases may get waterlogged or stuck to a film of condensation. Once a larva abandons its case, it is unlikely to make another or to recommence feeding. All species should therefore be kept outdoors until emergence is expected, either in sleeves or on potted plants. Many species appear to need sunshine to make them pupate and dew or rain is essential to sleeved larvae to prevent desiccation during dormant phases. Seed-feeders may be reared on picked seedheads placed in pots containing soil and litter, provided these are ventilated.

446(984). **Coleophora leucapennella** (Hübn.)

O. Viviparous.

L. 7-4. On the seeds of *Lychnis flos-cuculi*, in a case formed of a seedhead; leaves the plant to hibernate.

P. 5.

I. End 5-mid 6. In woodland rides, not rushy fields.

The *serratella* group. The first cases of no. 447 and nos 449-452 are pistol-shaped, bent almost at a right-angle with a silken keel subtended across the angle. In spring they are extended by the addition of narrow rings of leaf-cuticle to the oral end.

447(974). **C. lutipennella** (Zell.)

O. On a leaf of *Quercus* spp.

L. 10-5. First in a minute mine, containing frass; from this it cuts a tiny case, feeding from which it peppers the leaf with scattered pinhead-sized mines. Overwinters in the axil of a bud. In spring, it makes blotch mines, first adding small increments to its autumnal case, then cutting a final case from the mined margin of a leaf. *Cf. C. flavipennella.*

P. 6. Commonly on the upperside of a leaf.

I. 6-7.

448(967). **C. gryphipennella** (Hübn.)

O. On *Rosa* spp.

L. 9-5. Mines the leaves. In autumn in one or more flat, serrated, bivalved cases, each of a mined leaf-margin; in spring in a more cylindrical case, which is strongly reinforced internally with silk.

P. 6. Usually on a stem.

I. 6-7. Sexually dichroic. Occasionally a pest in gardens.

449(974). **C. flavipennella** (Dup.)

O. Beneath a leaf of *Quercus* spp.

L. 10-5. Life history not yet distinguished from that of *C. lutipennella*.

P. 6.

I. 6-7.

450(973). **C. serratella** (Linn.) (*fuscedinella* Zell.)

O. Beneath a leaf of *Alnus glutinosa, Betula, Corylus* or *Ulmus*; occasionally on *Sorbus aucuparia, S. aria* or *Carpinus*.

L. 9-5. Mines the leaves. Occasionally on herbaceous plants beneath the normal foodplants. Often a pest of birch saplings and elm hedges.

P. 6. Commonly on the upperside of a leaf of the foodplant or adjacent herbage.

I. 6-7.

451(965 partim) **C. coracipennella** (Hübn.) (*nigricella* sensu Ford, partim)

O. Chiefly on *Malus* in south-east England, occasionally on *Crataegus*.

L. 9-5. Makes blotch mines in the leaves from a dark, red-brown case fixed nearly upright.

P. 6.
I. Mid 6- mid 7. Now scarce.

452(965 partim). **C. cerasivorella** Packard (*nigricella* sensu Ford, partim)
O. Beneath a leaf of *Crataegus* or *Prunus* spp.; also on *Malus* or *Pyrus*.
L. 9-5. Not yet distinguished from *C. coracipennella*. Cases made from leaves of *Crataegus* and *Prunus* are darker. Not confirmed from *Sorbus* (cf. *C. serratella*).
P. 6.
I. Mid 6- mid 7. Abundant.

453(976 partim). **C. milvipennis** Zell. (*limosipennella* sensu Ford, partim)
O. Beneath a leaf of *Betula*.
L. ?7-4. Mines the leaves. Makes at least two cases, the final one 8-9 mm long; these are light brown, flat at first, toothed along the back, bivalved, straight or slightly curved following the leaf-margin from which they have been cut. Overwinters full-fed from mid-9.
P. 5.
I. Late 5- mid 6. Found at least as far north as the Midlands.

454(976 partim). **C. alnifoliae** Barasch (*limosipennella* sensu Ford, partim)
O. Beneath a leaf of *Alnus glutinosa* or *Betula* spp.
L. 9-6. Mines the leaves. First case 2 mm, second 3.5 mm long; hibernates in either, but may overwinter twice. Spring case 10-11 mm long, resembling that of *C. milvipennis*. Mine larger with a smaller hole than that of *C. serratella* feeding concurrently.
P. 6-7.
I. 7-8. Found only in south-east England.

455(976 partim). **C. limosipennella** (Dup.)
O. Beneath a leaf of *Ulmus*.
L. ?-mid 7. Mines the leaves. Last case *alnifoliae*-type, dark brown, bivalved, expanded at the tip, 9-11.5 mm long, m.o. 0°-20°, and held clear of the leaf by a distinct neck. Small larvae may be expected to occur in autumn, but have not been found. Larval stage may last two years.
P. 6.
I. End 6- mid 7. The year after feeding is completed. South-east England.

456(978 partim). **C. badiipennella** (Dup.)
O. Laid singly beneath a leaf of *Ulmus* in the angle between the midrib and a lateral vein.
L. 9-5. First in a straight, linear mine containing frass, from the end of which the first case, an ellipse measuring 3.0-4.0 x 1.0-1.5 mm, is excised. The final case, usually cut from the basal margin of a leaf, is bivalved, measures 5.0-8.0 x 1.5-3.0 mm, m.o. 10°-20°, held clear of the leaf by a neck. Larvae may overwinter in the first case or complete their growth in the autumn.
P. 6-7. Attached to a leaf or twig.

456A(978 partim). **C. adjectella** Herr. - Schäff. *Ent. Rec.*, in press (?1979)
O. On a leaf of *Prunus spinosa.*
L. 9-5. Mines the leaves. Case similar to that of *C. milvipennis* but smaller (*c.* 8 mm long); m.o. 20 , giving a more recumbent attitude than that of *C. milvipennis.* Overwinters half-fed.
P. 6.
I. 7.

457(–). **C. hydrolapathella** Her. *Ent.Rec.* **90**:15-17(1978)
O. On a floret of *Rumex hydrolapathum*.
L. 9-6. In a dark purple-brown, silken, trivalved case with angled neck, boring into the seeds of the foodplant. Spends the winter in the case spun to a plant stem.
P. 6-7.
I. 7-8. Occurs in the Norfolk Broads; will probably be found beside rivers elsewhere. Flies for two hours before dark.

458(996). **C. siccifolia** Staint.
O. On a leaf of *Betula* spp., *Crataegus* or *Malus*.
L. 9-8, but timing erratic and incompletely observed, the cycle possibly taking two years. Cases 2 and 3 are formed by cutting out a very large mine, rolling over one edge as a wrapping for a silken tube and leaving the rest as an appendage, the whole looking like a withered leaf. Other mines are also very large. Overwinters in case 1, about 3.5 mm long.
P.
I. 6.

459(977). **C. trigeminella** Fuchs
O. On *Crataegus; *Sorbus.*
L. ?9-early 5. In a very dark silken case, 5-6 mm long, bivalved when small and not always becoming trivalved, m.o.0° with a distinct neck. Mines leaves and bores into flower-buds from mid-April as the shoots expand.
P. 5.
I. 6. So far taken only in south-east England.

460(964). **C. fuscocuprella** Herr.-Schäff.
O. On *Corylus* or *Betula* spp.
L. 8-4. On hazel usually mining a lower terminal leaf, peppering it with numerous small blotches. Case ochreous brown, pistol-shaped, successively larger leaf-fragments making the coil almost globular. Does not feed again after hibernation.
P. 4-5. Off the bush.
I. 5-6.

461(972). **C. viminetella** Zell.
O. On *Salix*, mainly *S. caprea, S. cinerea, S. aurita* or *S. repens*, or *Myrica*.
L. 9-5. Mines the leaves. Case of mined leaf finally tricolorous, comprising a dark overwintered tip, a red-brown central piece and a greenish ochreous oral portion, stacked one behind the other with oblique junctions and slightly irregular outline. Old mines are rufous.
P. 6.
I. 6-7.

462(1040). **C. idaeella** Hofm. (*vacciniella* sensu Ford) *Ent.Gaz.* **17**:221(1966)
O. On *Vaccinium vitis-idaea*.
L. 8-4. Mines the leaves. Case formed of three pieces of mined leaf stacked end to end as in *C. viminetella*.
P. 5.
I. 6-7.

463(969). **C. vitisella** Gregs.
O. 7. Under a leaf of *Vaccinium vitis-idaea*.
L. 8-5. Lives two years, diapausing in spring and mining only mature leaves. Case made of rings of leaf added at the oral end; first-year case 4.0 x 4.0 mm, its tip curved back nearly to the leaf; second-year case 6.0-7.0 mm long, erect and pistol-shaped.
P. 5. On the upperside of a leaf.
I. 6-7.

464(970). **C. glitzella** Hofm.
O. On a leaf of *Vaccinium vitis-idaea*.
L. In the first autumn in a nepticulid-like mine with a broad frass-track, in spring making a longitudinal slit at the end. Then blotch mines with irregularly laid frass are made until early summer. The larva then diapauses until the next spring. A broad, frass-filled mine then expands into a blotch at the leaf-margin, where a case is cut out. Normal coleophorid fleck mines are finally made.
P. 5.
I. 6.

465(1005). **C. arctostaphyli** Meder (*marginatella* sensu auctt.)
O. Beneath a leaf of *Arctostaphylos*.
L. ?-4. In a narrow gallery mine from autumn of the first year until spring of the second year, then making an oval case and mining leaves. Aestivates late 6-early 8. Final case the following April resembles a dried leaflet.
P. 5.
I. 6.

466(963, 960A). **C. violacea** (Ström) (*paripennella* sensu auctt.; *hornigi* Toll; *albicornuella* Bradley)
O. Beneath a leaf of one of various shrubs, such as *Corylus, Malus, Prunus, Rosa, Rubus, Betula* or *Ulmus*.
L. 7-4. Mines the leaves. Case brown, with anterior rings of projecting leaf-fragments, the mines often having a hole in both cuticles when these parts have been used to enlarge the case; the case blackens and the rings erode in winter; m.o. 0°. Full-fed in autumn.
P. 4-5. Near the ground.
I. 5-6.

467(982). **C. juncicolella** Staint.
O. On *Calluna* or *Erica*.
L. 9-4. Case very small, made from several leaves arranged to resemble a fragment of a shoot of *Calluna*. Best obtained by sweeping.

P. 5. Attached to a stem.
I. 6-7. Will fly by day. A very small species, easily overlooked amongst *Juncus*-feeders.

468(968). C. **orbitella** Zell.
O. Beneath a leaf of *Betula* spp.; **Alnus*.
L. 8-4. Mines the leaves. Case made of a single piece of leaf, often with a mid-ventral projection; dark brown, becoming darker grey over winter. Overwinters full-fed.
P. 4-5.
I. 5-6.

469(971). C. **binderella** (Koll.)
O. Beneath a leaf of *Alnus glutinosa, Betula* or *Corylus*.
L. 9-5. Mining the leaves. Case somewhat laterally compressed, usually made of three leaf-fragments, the freshest one brighter-coloured than the others and protruding untidily from the outline of the case.
P. 6.
I. 6-7.

470(962). C. **potentillae** Elisha
O. Beneath a leaf of herbaceous Rosaceae or *Rubus;* locally on *Rosa* or *Helianthemum*.
L. 9-5. Mines the leaves. Case resembles that of *C. violacea*. Usually overwinters full-fed but larvae in small cases may feed until June of their second year; bienniel in Co. Clare.
P. 5. In Co. Clare on stems of *Rubus saxatilis*.
I. 6.

471(961). C. **ahenella** Hein.
O. Beneath a leaf of *Rhamnus*; also *Viburnum* or *Swida*.
L. Bienniel; feeding periods 8-9 in first year, 5-6 (in small case) and 8-9 in second year. Case resembles that of *C. violacea*.
P. 5.
I. 6.

472(959). C. **albitarsella** Zell.
O. Beneath a leaf of *Calamintha* spp., *Clinopodium, Glechoma, Mentha* spp. or *Origanum*.
L. 9-5. Mines the leaves, making blotches which are conspicuously white when fresh. Case blackish brown, slightly curved, slender, with keeled tip, smoothly made of about 10 cylindrical sections of leaf.
P. 5. Near ground level.
I. 6-7.

473(958). C. **trifolii** (Curt.) (*frischella* sensu auctt., partim)
O. 7. On a flower of *Melilotus altissima* or *M. officinalis*.
L. 8-5. Eats the seeds. Case made of a young seedpod, then of up to three joined like beads; in final instar makes a trivalved, black case from a mature pod.
P. 6. At ground level.
I. 7. Flies by day.

474(–). **C. frischella** (Linn.) *(alcyonipennella* sensu auctt.) *Ent.Gaz.* **27**: 145-147(1976)

O. — *On a floret of *Trifolium repens*.

L. — *Bores through the florets and feeds on the green seeds. Case formed in 4th instar, *c.* 4 mm long, made of a withered standard petal lined with silk with trivalved apex.

P. — *In the feeding place or on the ground.

I. End 5-6; 7-9. — At least two broods: the only multivoltine British coleophorid. Mainly diurnal. Submaritime in southern Britain.

475(956). **C. spissicornis** (Haw.) *N.Z.Jl agric.Res.* **1**:240-248(1958)

O. — On a floret of *Trifolium repens*.

L. 8-5. — Case and habits scarcely distinguishable from those of *C. frischella*.

P. 5. — On the ground.

I. 6-7. — Diurnal.

476(957). **C. deauratella** L. & Z.

O. — On a floret of *Trifolium pratense*.

L. 8-9. — Bores through the florets and feeds on the green seeds. Case formed in the 4th instar, *c.* 6 mm long, resembling that of *C. frischella* but made more conspicuous by the dark brown trivalve projecting beyond the covering petal.

P. — In the seedhead or on the ground.

I. 6-7. — Diurnal.

477(–). **C. fuscicornis** Zell. *Ent.Rec.* **87**:259-266(1975)

O. 6. — On a flower of *Vicia tetrasperma*.

L. 7-5. — In a pod of the foodplant, later using this as a case to attack the seeds in further pods. The yellow-brown case shows up conspicuously when attached to green seedpods. For overwintering, a trivalved silk lining is formed entirely within the pod.

P. 5. — Near the ground.

I. End 5 - early 6. — Diurnal. Essex coast; may spread to other submaritime localities.

478(1013). **C. conyzae** Zell.

O. — Beneath a leaf of *Inula conyza* or *Pulicaria dysenterica*.

L. 9-5. — Mines the leaves. At least two cases are made, each from a mined portion of leaf; they are brown, hairy, bivalved and flat at first. Empty cases often remain fixed to the end of a cut-out mine.

P. 6. — Attached to the stem of the foodplant.

I. 7.

479(1012). **C. lineolea** (Haw.) *(crocogrammos* Zell.)

O. — Beneath a leaf of *Ballota nigra, Stachys lanata, S. sylvatica, Betonica officinalis* or *Marrubium vulgare*, against a vein.

L. 9-5. — Mines the leaves. Case broad, flattened and pointed posteriorly, assembled from transversely placed leaf-mines, colour and hairiness depending on the foodplant used.

P. 6. — Attached to the stem of the foodplant low down.

I. 7.

480(983). **C. hemerobiella** (Scop.)
O. Beneath a leaf of *Crataegus, Malus, Prunus avium, P. domestica* or *Pyrus; *Sorbus.*
L. 9-5. Bienniel, mining the leaves. The first case is made of leaf-cuticle, curved to half a toroid, without the silk ventral keel of the *serratella* group; in the second spring, a nearly erect, dark brown case, 8-10 mm long, is cut from a leaf-margin.
P. 6. Attached to a twig.
I. 7.

481(980). **C. lithargyrinella** Zell. (*olivaceella* Staint.)
O. On *Stellaria holostea.*
L. 9-5. Mines the leaves, making white blotches often late into autumn and in spring and occurring in small numbers, often amongst *C. solitariella.* Case at first formed of a leaf-mine, later extended in ochreous silk; m.o. about 20 ; a dorsal keel, anteriorly single but posteriorly double, extending to the two dorsal elements of the apical trivalve.
P. 6. Attached to a stem low down.
I. 6-7.

482(979). **C. solitariella** Zell.
O. On the upperside of a leaf of *Stellaria holostea.*
L. 9-5. Mines the leaves, making white blotches similar to those of *C. lithargyrinella.* In autumn a 3-4 mm case is cut from the apex of a short leaf-mine and is immediately extended with silk, but the anterior dorsal keel on this portion of the whitish case is not extended backwards when it is again enlarged in spring; m.o. greater than 45°. Often in dense colonies.
P. 6. On a stem low down.
I. 7.

483(981). **C. laricella** (Hübn.)
O. 6-7. Laid singly on a needle of *Larix*; hatches in 10-14 days.
L. 7-5. In the first two instars mines the needles. A case is formed of part of a needle in mid-September, mining continuing for about two weeks. Hibernation occurs from October to bud-burst in April. The nearly white, erect, stumpy case is enlarged by silken gussets.
P. 5-6. Attached to a leaf or twig.
I. 6-7.

484(985). **C. wockeella** Zell.
O. On *Betonica officinalis.*
L. 10-5. Mines the leaves. Case dark brown, long, rather flat, with hooked posterior end; more easily found in autumn.
P. 5-6. Attached to the stem of the foodplant or to adjacent herbage.
I. 6-7.

485(987). **C. chalcogrammella** Zell.
O. On a leaf of *Cerastium arvense.*
L. 10-5. Mines the leaves from a case constructed as that of *C. lithargyrinella*, but slender, more tapered posteriorly and orange-brown; ventral keel sharp, dorsal keels weak.

P. 5-6. Low down.
I. 6-7. On calcareous sandy soils.

486(988). **C. tricolor** Wals.
O. 8. On a flower of *Acinos*.
L. 8-5. At first on seeds of *Acinos*, making a case from a seedhead. Later, often from autumn, mining the leaves of grasses and cutting a case from a mine.
P. 6. Near the ground.
I. 7-8. Diurnal; feeds on the flowers of *Acinos*. On open plant communities on Breck and chalk soils.

487(989). **C. lixella** Zell.
O. 7. On a flower of *Thymus* spp.
L. 8-5. On the seeds of *Thymus*, using a seedhead as a case. In spring, mining leaves of grasses and cutting a case from a mine.
P. 6. Near the ground.
I. 7. Diurnal.

488(990). **C. ochrea** (Haw.)
O. On *Helianthemum chamaecistus*.
L. 10-5. Mines the leaves. Case of leaf-pieces: the overwintered part is grey, the fresh part ochreous. This large species is partial to plants overhanging ledges.
P. 6.
I. 7-8.

489(991). **C. albidella** ([D. & S.])
O. On *Salix caprea, S. cinerea* or *S. aurita*.
L. 9-5. In spring burrows into expanding buds, then skeletonizes young leaves. The 7 mm-long, black pistol-case is scattered with white, woolly leaf-hairs.
P. 6-7. On a stem or the upperside of a leaf.
I. 7-8.

490(992). **C. anatipennella** (Hübn.)
O. Beneath a leaf of *Crataegus, Malus* or *Prunus*, especially *P. spinosa*; occasionally *Sorbus*.
L. 8-5. In autumn, spins a minute black case from frass piled above a mine, then makes a patch of tiny mines on one leaf. In spring, at first mines, but after expanding its case eats out holes in the leaves. Case upright, black, pistol-shaped, expanded posteriorly.
P. 6. On a stem or the upperside of a leaf.
I. 6-7.

491(996). **C. currucipennella** Zell.
O. On *Carpinus, Quercus* spp. or *Salix* spp.; *other hardwoods.
L. ?-early 6. Skeletonizes oak and makes large holes in other leaves. Case black, silken, pistol-shaped; dorso-lateral flaps at the position of the mouth of the overwintered portion; successively smaller dorso-lateral projections at the apices of further additions to the case; posterior to a lobed ventral keel, a small ventro-lateral flap conceals the constricted section anterior to the narrow apical 'handle'.
P. 6. On a twig or the upperside of a leaf.
I. 6-7.

492(993). **C. ardeaepennella** Scott
O. Beneath a leaf of *Quercus* spp.
L. 9-6. Habits as for *C. anatipennella*. Case little expanded below, the stoutest part not reaching the outline of the 'handle', m.o. 60-80°.
P. 6-7. Attached to a leaf or twig.
I. 7-8.

493(994). **C. ibipennella** Zell. (*betulella* sensu auctt.)
O. Beneath a leaf of *Betula* spp.
L. 9-5. Eats out holes in or feeds at the edges of leaves in spring. Case black, pistol-shaped, m.o. 30-60°; ventral keel strongly swollen; a line from the oral end along the keel passes through the tip of the apical 'handle'.
P. 6-7. On a leaf or twig.
I. 7. Has a yellow-veined form resembling *C. currucipennella*.

494(995). **C. palliatella** (Zinck.).
O. On *Quercus* spp.
L. ?-5. Chews the edges of young leaves like a 'macro' larva in spring. Case large, black, pistol-shaped, almost enclosed by translucent, scaly flaps (the 'pallium').
P. 6-7. On the upperside of a leaf of any tree or bush.
I. 7.

495(997). **C. vibicella** (Hübn.)
O. On *Genista tinctoria*.
L. 9-6. Consumes leaves from above in spring, but rests on stems. Case made of silk, elongate, black, pistol-shaped.
P. 6-7. On a stem.
I. 7. Very locally abundant.

496(998). **C. conspicuella** Zell.
O. On *Centaurea*.
L. 8-5. Mines the leaves. Case made of silk, black with a clear sheen, slender, bowed, thinly keeled, up to 15 mm long. Larva brown.
P. 6.
I. End 6-early 7. Abundant very locally on coastal sites; formerly on Surrey downs.

497(999). **C. vibicigerella** Zell.
O. On *Artemisia maritima; *Achillea*.
L. 9-5. Mines the leaves. Case black, recumbent, bellied beneath a slender tube.
P. 6.
I. End 6 - early 7. Not observed in recent years.

498(1000). **C. pyrrhulipennella** Zell.
O. On *Calluna* or *Erica*.
L. 9-5. Mines the leaflets. Case of black silk, up to 9 mm long, slender, but swollen by a thin central keel.
P. 6. Attached to a stem or the tip of a shoot in full sunshine.
I. 7.

499(–). **C. serpylletorum** Her. *Proc. S. Lond. ent. nat. Hist. Soc.* **1966**(3): 73 and pl.2,fig.15.
O. On *Thymus* spp.
L. 9-6. Mines the leaves, especially from the tips. Case made of about five pieces of mine added irregularly crosswise.
P. 6.
I. 7. Western sea-cliff species.

500(1001). **C. vulnerariae** Zell.
O. 7. On a flower of *Anthyllis vulneraria*.
L. 7-?. The young larva bores from seedpod to seedpod. The case, made of a calyx, sometimes with the floret attached, is probably made in the last instar.
P.
I. 6-7.

501(1007). **C. albicosta** (Haw.)
O. On a flower of *Ulex europaeus*.
L. 7-4. Clears a pod of green seeds, making little silk, then emerges to make a case of the lobe of the calyx. Other pods are bored into from the side until the larva is mature. Overwinters full-fed.
P. 5. On a stem.
I. 5-6. Diurnal.

502(1002). **C. saturatella** Staint.
O. On *Sarothamnus* or *Genista tinctoria*.
L. ?-6. Mines the leaves, especially from the tip. Overwinters in a minute case of leaf-fragments; in spring, cuts large pieces of leaf to add across its broad, flat, herring-bone-patterned case. Case yellow-brown on *Genista*, dark brown, aging to dark grey, on *Sarothamnus*.
P. 6-7. On a stem of the foodplant, often high up in sunshine.
I. 7. The species appears to change its ground, old cases often being found where fresh ones are not to be seen.

503(1003). **C. genistae** Staint.
O. On *Genista anglica*.
L. 9-5. Mines the leaves, blanching them. Case light and dark brown, rather flattened and made of leaf-fragments in a herring-bone-pattern, resembling a spikelet of grass such as *Bromus*.
P. 5-6. Attached to a stem of the foodplant.
I. 6-7. Often a pest of its foodplant.

504(1004). **C. discordella** Zell.
O. On *Lotus corniculatus*.
L. 9-5. Mines the leaves, which quickly shrivel. Case blackish brown, somewhat flattened, curved posteriorly, made from broad, transverse bands of leaf-cuticle.
P. 6. Low down.
I. 7.

505(1006). **C. niveicostella** Zell.
O. On *Thymus* spp.
L. 9-5. Mines the leaves; mines are few and quickly darken, be-

coming inconspicuous. Case dark brown, about 7 mm long, slender, made from transverse portions of leaf expanded with silk.

P. 5-6. In clumps of the foodplant.

I. 6-7.

506(1008). **C. onosmella** (Brahm)

O. On a leaf of *Echium* or *Anchusa*.

L. 9-5. Mines the leaves from the underside or the edge. Case broad, flattened, blackish brown but superficially grey from the hair on the transversely arranged leaf-bands.

P. 5-6. Attached to a stem.

I. 6-7.

507(1009, 1017). **C. silenella** Herr.-Schäff. (*nutantella* Mühlig & Frey; *graminicolella* Hein.)

O. On a flowerhead of *Silene vulgaris* or *S. nutans*.

L. 8-5. On the seeds, extruding a white, cylindrical silken case from the apex of the ovary. Later the case is attached to other seed-capsules, often at the side.

P. 5-6. Away from the feeding site, where empty cases often remain.

I. 6-7.

508(1010). **C. otitae** Zell.

O. On a leaf of *Silene nutans*.

L. 5-7. Mines beneath the leaves near the ground. Case of grey-white silk, decorated with longitudinal stripes of fuscous particles.

P. 7. On the ground.

I. 8. Abundant at Dungeness; not yet found elsewhere.

509(1011A). **C. lassella** Staud. *Ent.Gaz.* **3**:190-191(1952)

O.

L.

P.

I. Mid 6.

510(1011). **C. striatipennella** Nyl. (*lineolea* sensu Ford)

O. On a flower of *Stellaria graminea*; also on *S. media* in Eire.

L. 8-5. In seedpods, extruding a yellow-grey, cylindrical case from the first pod attacked.

P. 6. Near ground level.

I. 7-8.

511(1014). **C. inulae** Wocke

O. 7. Beneath a leaf of *Pulicaria dysenterica* or *Inula conyza*.

L. Bienniel. Overwinters first in a small case and generally feeds up by the following July, though some further feeding may take place after the second winter. Case up to 15 mm long, m.o. necked at 0°. The antero-dorsal first-year portion is grey, the remainder of the cigarette-like silk tube passing through ochreous to creamy white.

P. 5-6. Near ground level.

I. 6-7.

512(1015 partim). **C. troglodytella** (Dup.)

O. Beneath a leaf of *Cirsium arvense, Eupatorium, Inula* or *Pulicaria*; probably also on the other plants listed for *C. trochilella*.

L. 9-early 6. Mines the leaves. Overwinters in a small case which persists as the antero-dorsal part of the completed silk tube; in spring the case is expanded beneath and behind so that its profile is somewhat convex and the length is 7-9 mm. In some years, colonies on *Eupatorium* on the coast extend their cases to 13-14 mm at maturity, producing a tapered section and more angled neck, thus resembling *C. inulae*; the larva may be distinguished from that species by the absence of an obvious dark sclerite situated postero-dorsally to the spiracle on abdominal segment I.

P. 6. Near the base of a stem.

I. 7-8.

513(1015 partim). **C. trochilella** (Dup.) (*troglodytella* sensu Ford, partim) *Ent.Gaz.* **13**:184(1962)

O. Beneath a leaf of *Artemisia vulgaris, Achillea, Eupatorium* or *Tanacetum*.

L. ?-6. Mines the leaves. Case not distinguished from that of *C. troglodytella*.

P. 6-7. Near the base of a stem.

I. 7-8.

514(–). **C. machinella** Bradley *Ent.Gaz.* **22**:24-25(1971)

O. Beneath a leaf of *Achillea ptarmica* or *Artemisia maritima*.

L. 9-6. Mines the leaves. Overwinters in a small case; mature case on *Achillea* of bright ochreous silk, 7 mm long, slim, distinctly necked, m.o. 60°.

P. 6.

I. 7.

515(–). **C. ramosella** Zell. *Ent.Gaz.* **11**:31-36(1960)

O. Beneath a leaf of *Solidago virgaurea*.

L. ?-6. Mines the leaves, the mines darkening rapidly. Case silken, cream to light ochreous brown, 7-9 mm long, slender to stout, m.o. 20°. In the Burren of Co. Clare, on the leathery basal leaves of plants growing in crevices of the limestone pavement.

P. 6. In the crevices.

I. 7.

516(1016). **C. peribenanderi** (Toll) (*therinella* sensu auctt.)

O. Beneath a leaf of *Cirsium arvense*; sometimes on *Arctium* spp.

L. 9-5. Mines the leaves. Case grey-white, silken, 11-12 mm long, slightly convex in profile, m.o. 45-60°. Usually full-fed in autumn, but occasionally feeds again in spring.

P. 5-6. Probably low down.

I. 6-7. Flies throughout both months.

517(960). **C. paripennella** Zell. (*alcyonipennella* sensu P. & M. et Ford)

O. Beneath a leaf of *Centaurea* or, occasionally, *Arctium* spp. or *Serratula*.

L. 9-6. Mines the leaves. Case nearly black, matt, not keeled, up to 8 mm long, m.o. 0-20°; there is a lateral gusset of pale silk during periods of expansion.
P. 6. On a stem or on the ground.
I. 6-7.

518(–). C. **therinella** Tengst. *Ent.Gaz.* **11**:35(1960)
O.
L.
P.
I. 6-7.

519(1018). C. **asteris** Mühl. (*tripoliella* Hodgk.)
O. On a flower of *Aster tripolium*.
L. 9-5. On the seeds; leaves the plant before the winter. Case dark brown, m.o. 90°.
P. 6.
I. 7-8.

520(1019). C. **argentula** (Steph.)
O. On a flower of *Achillea millefolium*.
L. 9-5. On the seeds in autumn. Case light brown, stout, rough-surfaced.
P. 6. Near the ground.
I. 7-8.

521(1020). C. **virgaureae** Staint.
O. On a flower of *Solidago virgaurea* or *Aster tripolium*.
L. 9-5. On the seeds. Case silken, white at first, darkening to brown; m.o. 45° or less. On *Solidago*, the case has a porcupine-like covering of pappus hairs attached near the front.
P. 6. At ground level.
I. 7-8. Flies in May in the Burren of Co. Clare.

Species feeding on seeds of Chenopodiaceae. These have been the subject of much misidentification. Neither larvae nor cases of some species have yet been distinguished. Mixed series will be bred from a single locality and foodplant. Some species have preferred hostplants, but most *Chenopodium* and *Atriplex* will be attacked; *Beta* is normally immune. All overwinter full-fed.

The figures of genitalia given by Patzak (*Beitr. Ent.* **24**:153-278(1974)) are the most correctly named. Pelham-Clinton (*Ent.* **92**:120-124(1959)) keys most of the species.

522(1023 partim). C. **benanderi** Kanerva (*flavaginella* sensu Ford et Pelham-Clinton; *annulatella* sensu auctt.)
O. On a floret of *Chenopodium* or *Atriplex*.
L. 9-6. On the seeds. Case short, pale brown studded with meal from the plant. Full-fed in September.
P. 6. On the ground.
I. 7-8. Ubiquitous: the commonest species inland.

523(1023 partim). **C. sternipennella** (Zett.) (*flavaginella* L. & Z.)
O. On a floret of *Chenopodium* or *Atriplex*, especially *C. album*.
L. 9-6. Not distinguished from *C. vestianella*.
P. 6. On the ground.
I. 7-8. Frequent on waste ground.

524(1024A). **C. adspersella** Benander
O. On *Atriplex, Halimione* or *Suaeda maritima*.
L. 9-5. On the seeds. The case, the largest of the group, is 8 mm long, swollen in the middle, with prominent valves, pale and little decorated. It is already large when *C. sternipennella* is minute.
P. 6. Low down.
I. Late 6. Salt-marshes, fens, Breckland and dry fields elsewhere.

525(–). **C. versurella** Zell. *Ent.* **92**:27-33(1959)
O. On *Chenopodium* or *Atriplex littoralis*.
L. 9-5. On the seeds. Case 5-6 mm long, stout, of grey silk with smoother dark grey longitudinal seams; m.o. 30°.
P. 5-6. Low down.
I. 6-8. On salt-marshes and waste ground.

526(1021). **C. squamosella** Staint. (*erigerella* Ford)
O. On a flower of *Erigeron acer*.
L. 9-5. On the seeds. Case slate-coloured, hidden by the pappus hairs.
P. 6-7. At ground level.
I. 8.

527(–). **C. pappiferella** Hofm. *Ent.Gaz.* **13**:171-174(1962)
O. On a flower of *Antennaria*.
L. 7-4. On the seeds, feeding concealed in the heads. Case starts as a flimsy white silken tube, becoming dark brown and decorated with pappus hairs.
P. 5.
I. End 5-early 6.

528(1025A). **C. granulatella** Zell.
This species appears to have been added to the British list through misidentification.

529(1024). **C. vestianella** (Linn.) (*laripennella* (Zett.); *annulatella* Nyl.)
O. On a floret of *Atriplex patula*.
L. 9-5. On the seeds.
P. 5-6. Amongst debris on the ground.
I. 6-7. Breckland and Lancashire sandhills.

530(1025). **C. atriplicis** Meyr.
O. On a floret of *Halimione, Suaeda maritima, Salicornia* or *Atriplex littoralis*.
L. 8-5. On the seeds. Case brown-grey.
P. 6. Attached to a stem.
I. End 6-early 7. On salt-marshes.

531(1026). C. **suaedivora** Meyr. (*moeniacella* sensu auctt.)
O. On *Suaeda*.
L. 7-5. Mining leaves and eating seeds. Case ochreous white with darker stripes.
P. 5-6. At ground level.
I. 6-7.

532(1027). C. **salinella** Staint.
O. On *Halimione* or *Atriplex littoralis*.
L. 9-6. On the seeds; on *Halimione* also mines the leaves. Case soft and pale, little decorated by granules, not much stouter in the middle; length 5-6 mm, m.o. 45°.
P. 6-7. Attached to a stem on *Halimione*.
I. 7-8.

533(1028). C. **artemisiella** Scott
O. On *Artemisia maritima*.
L. 8-6. At first mining the leaves, later on the seeds in a soft, ochreous grey silken case. Full-fed in September.
P. 6-7. Near the ground.
I. 7-8.

534(1029). C. **artemisicolella** Bruand
O. On a flower of *Artemisia vulgaris*.
L. 8-6. On the seeds. Case formed of an empty seedhead; this is stuck to the side of other seedheads as they are attacked, leaving a tell-tale hole after removal. Full-fed in September.
P. 6-7. Attached to a stem low down.
I. 7-8.

535(1030). C. **murinipennella** (Dup.)
O. In a flower of *Luzula campestris* or *L. multiflora*.
L. 7-4. In the seed-capsule; later on the ground on ripe fallen seeds until September. Case of brownish ochreous silk, decorated with seed debris anteriorly, formed by the young larva in the seedhead and enlarged whilst it feeds on the ground.
P. 4-5.
I. 5-early 6.

536(1031). C. **antennariella** Herr.-Schäff.
O. *On a flower of *Luzula pilosa*.
L. *On the seeds. Case a floret, with the tubular, silken tip whitish grey in spring; perhaps behaves as *C. murinipennella*.
P. 4. Often at the base of a tree-trunk in a sunny situation.
I. End 4-mid 5.

537(1032). C. **sylvaticella** Wood
O. In a floret of *Luzula sylvatica*.
L. Bienniel. On the seeds. Case formed of a floret surrounding an ochreous silken tube with a distinct neck. Larvae feed in late 7-10, some again on new panicles in early 5, complete their cases in early 6 and leave the flowers in 7-8; others do not feed in their second year.

P. 4-5. On woodland litter.
I. 4-5 or early 6. In sheltered woodlands.

538(1033). C. **taeniipennella** Herr.-Schäff. (*galactaula* Meyr.)
O. On a floret of *Juncus articulatus* or *J. acutiflorus*, overspill of dense populations on *J. effusus*, occasionally on *J. inflexus* and *J. subuliflorus*.
L. 9-5. On the seeds. Case made in the third instar by cutting off a complete seedhead, still closed at the top; the seedhead remains girdling the base of the creamy white silk case; length about 5 mm. Feeds until October; some larvae overwinter on the foodplant.
P. 5-6.
I. Mid 6-7.

539(1035). C. **glaucicolella** Wood
O. On a floret of *Juncus* spp.; inland mainly on *J. inflexus* because of the more persistent seeds, but also on *J. subuliflorus, J. effusus, J. articulatus* and *J. acutiflorus*; on saltmarshes abundant on *J. maritimus* and *J. gerardii*.
L. 9-5. On the seeds. Case of white silk, tubular, like that of *C. alticolella* but formed in the 2nd instar and so more often seen small. Hibernates at a variable size on or beneath the plants and continues feeding in early spring. Larva with dorsal sclerites of metathorax faintly coloured.
P. 6-7.
I. 7-8. Much later than *C. alticolella*.

540(1034). C. **tamesis** Waters *Ent. Rec.* **82**:1-2(1970).
O. On a floret of *Juncus articulatus*.
L. 9-5. On the seeds. Case a muddy-fuscous silken tube covered at the base by a complete floret; length 6-7 mm. Full-fed by early September.
P. 5-6.
I. Mid 6 - mid 7 or early 8.

541(1036). C. **alticolella** Zell. (*caespititiella* sensu auctt.) *Ent.* **88**:273-277 (1955); *Trans. Soc. Brit. Ent.* **13**:1-16(1958)
O. 6-8. Laid on the inner surface of a perianth segment of *Juncus*, or in the angle between two adjacent flowers; mainly on *J. squarrosus* and *J. effusus*, but also on *J. inflexus, J. subuliflorus, J. articulatus, J. acutus, J. compressus*.
L. 7-5. When young, on the green seeds within a capsule. Makes a white, tubular silken case with adherent seed debris in 3rd or 4th instar. Hibernates in litter below plants and only feeds again in spring in case of starvation. Larva with no dorsal plates visible on metathorax.
P. Mid 5-early 6.
I. 6-7. Rests on rushes and easily disturbed by day.

542(1037). C. **maritimella** Newman (*obtusella* Staint.)
O. On a flower of *Juncus maritimus*.

L. 9-5. On the seeds. Case completely enclosed in a seedhead and hard to find; best taken by searching for small holes in the sides of florets, which indicate the feeding of larvae.
P. 5-6. Attached to a stem of the foodplant.
I. 6-7.

543(1038). **C. adjunctella** Hodgk.
O. On a floret of *Juncus gerardii*.
L. 8-5. On the seeds. Case completely enclosed in a floret. Early feeding detectable from boreholes in the side of seedheads; these are also made by the more plentiful *C. glaucicolella*, but by the time this species feeds, most *adjunctella* larvae have left the plants for hibernation. Collect by sampling.
P. 5-6. At ground level.
I. 6-7.

544(1039). **C. caespititiella** Zell. (*agrammella* Wood)
O. On a floret of *Juncus subuliflorus*; also on *J. effusus* and *J. articulatus*; very rarely on *J. inflexus*.
L. 8-5. On the seeds. Case silken, tubular, very small, grey, sometimes almost black, longitudinally striped with vegetable particles.
P. 5-6.
I. 6. In woodlands.

545(986). **C. salicorniae** Wocke
O. On *Salicornia*, typically on small plants colonizing well-drained bare patches in the mixed vegetation zone.
L. 9-10. Boring internally and eating the seeds, finally cutting off the tip of the stem as a case, attaching it to another stem like a branch to continue feeding. Pale 'sawdust' frass in the stub end of the first stem is a tell-tale. When full-fed, it walks or falls to the mud and abandons its case.
P. 6-7. In a parchment cocoon within a silk tube spun downwards in the mud. Can pupate in the case in captivity. Pupa normal (*cf* no. 546).
I. Mid 7-8. Occasionally recorded inland, so presumably disperses long distances.

546(1040A). **C. clypeiferella** Hofm.
O. On a floret of *Chenopodium album*.
L. 8-6. Spins a feeble tube through a group of seedheads, cuts these off as a crude case and continues to feed on other seeds; the case shows up as the component florets turn pale brown. When full-fed in September, falls to the ground and abandons the case.
P. 6-7. In a tough, brown, sealed cocoon in loose soil. Pupa membranous, sloughed like a larval skin upon emergence.
I. 7-8. Commonest mid 7. Cuts its way out of the cocoon by using the spines on abdominal segment 1, then pushes its way out of the soil before expanding its wings. Waste ground, especially London suburbs and Breckland.

ELACHISTIDAE

547(850). **Perittia obscurepunctella** (Staint.) (*obscuripunctella* missp.)
O. On *Lonicera periclymenum*.
L. 6-8. In a large, inflated blotch with a central, reddish-brown spot.
P. 9-4. In a cocoon in the bark of the foodplant.
I. 4-5. Flies in afternoon sunshine.

548(850A). **Mendesia farinella** (Thunb.) *Microscope* **7**:299-303(1950)
O.
L. Possibly on Boraginaceae.
P.
I. 5-6. One specimen was taken at Dover, Kent in 1897.

549(851). **Stephensia brunnichella** (Linn.) (*brunnichiella* missp.)
O. On *Clinopodium*, on the underside of a leaf, near the midrib.
L. 4;7. In a gallery leading to a brownish blotch.
P. 4-5;7-8. In a cocoon under a leaf, usually the leaf opposite the mined leaf.
I. 5;8. Occurs mainly on chalk soil.

Elachista, **Biselachista** and **Cosmiotes**. The larvae of all species mine leaves of Gramineae, Cyperaceae or Juncaceae. Most move to fresh leaves naturally or at least in captivity, but a few will not. Moths of all species fly on calm evenings and sometimes also at night, but males may sometimes be seen in the greatest numbers on wing at sunrise in calm, clear conditions.

550(854). **Elachista regificella** (Sirc.) (*magnificella* (Zell.) nec (Dup.))
O. On *Luzula* spp., especially *L. sylvatica* and *L. pilosa*.
L. 9-6. In a long, inflated mine, often changing to a second leaf; develops slowly.
P. 6-7. At the base of a leaf, not that which was mined; on the underside.
I. 7.

551(855). **E. gleichenella** (Fabr.)
O. On *Carex* spp., especially *C. flacca, C. echinata* or *C. paniculata*, or on *Luzula pilosa*.
L. 9-5. In a mine as broad as the leaf, directed towards the base. Larva whitish grey, head dark brown, prothoracic plate black.
P. 5-6. On the ground, between loosely spun leaves.
I. 6-8.

552(876). **E. biatomella** (Staint.)
O. On *Carex flacca*, in the angle of the leaf.
L. 9-4;7. In a gallery, directed first towards the tip of the leaf, later towards the base. Larva yellowish, head dark brown, prothoracic plate light brown.
P. 5;7-8. In the angle of a lower leaf.
I. 5-6;8.

553(863). E. **poae** Staint.
O. On *Glyceria maxima*, at the tip of a leaf.
L. 4;7. In a mine paler than the colour of the leaf, leading downwards into the stem.
P. 4-5;7-8. On an unmined leaf, under a slight cocoon.
I. 5;8.

554(860,858). E. **atricomella** Staint. (*holdenella* Staint.)
O. On *Dactylis*.
L. 3-5. In successive linear, whitish mines with a fine line of greyish frass, the last mine leading down into the stem or even the roots. Larva pale yellow, head and divided prothoracic plate yellowish brown.
P. 5-6. On the stem, supported by a girdle.
I. 6-8.

555(862). E. **kilmunella** Staint.
O. Possibly on *Carex* spp. or on *Eriophorum* spp.
L.
P. Under a girdle.
I. 6-7. On moors; flies about 1700 hrs.

556(861). E. **alpinella** Staint.
O. On *Carex* spp., especially *C. acutiformis*.
L. 9-7. In a long, narrow gallery, extending downwards into the stem.
P. 7. On a stem or leaf, under a girdle.
I. 7-9. May be obtained by day by sweeping the foodplant.

557(859). E. **luticomella** Zell.
O. On *Dactylis*.
L. 11-5. Mine similar to that of *E. atricomella*; larva also similar, but slightly darker yellow.
P. 5-6. On detritus or at the base of a leaf of the foodplant.
I. 6-8.

558(857). E. **albifrontella** (Hübn.)
O. On one of a wide range of Gramineae, especially *Dactylis, Holcus* spp. and *Deschampsia cespitosa*.
L. 9-5. In a broad gallery extending from the tip, containing narrow lines of greenish frass; mines several leaves. Larva whitish yellow, head pale brown, prothoracic plate divided into several brown spots.
P. 5-6. On detritus.
I. 6-7.

559(856). E. **apicipunctella** Staint.
O. On one of a number of Gramineae, especially *Dactylis, Festuca* spp. and *Holcus* spp.
L. 10-4. In a gallery extending from the tip of the leaf, at first narrow, then widening gradually; the edges are irregular and patches of uneaten parenchyma are left in the mine; sometimes there is more than one larva in a mine; hibernates full-fed from December in the mine. Larva yellowish, later whitish; prothoracic plate consisting of two elongate, L-shaped marks.

P. 4-5. On a stone or stem.
I. 5-6. Rests on tree-trunks.

560(865). **E. subnigrella** Dougl.
O. On *Bromus erectus*.
L. 4-5;7. In a flat, narrow mine with a nearly continuous line of frass; the mine is at first yellowish green, but later turns purplish. Larva pale yellow; head and prothoracic plate brown.
P. 5;7-8. Under a girdle at the point where the mined leaf branches from the stem, or on adjacent plants.
I. 5-6;8.

561(–). **E. orstadii** Palm *The Elachistidae (Lepidoptera) of Fennoscandia and Denmark*:88(1977)
O.
L. Early stages unknown.
P.
I. 5-6. Has been taken in the extreme south-east of Scotland.

562(–). **E. pomerana** Frey *Antenna* 2:22(1978)
O. *On *Poa pratensis, Phalaris arundinacea, Glyceria fluitans, Calamagrostis* or *Avena* spp.
L. *4-5. In a broad gallery extending from the tip of the leaf; the frass is at first dispersed, but later in a narrow line, changing from dark green to black. Larva yellowish, flecked with white, gut dark green; head and prothoracic plate light brown.
P. *5. Under a dense web on a leaf, often close to the surface of the water.
I. 6-7. Frequents fenland.

563(864). **E. humilis** Zell. (*perplexella* Staint.)
O. On *Deschampsia cespitosa*.
L. 4-5. In a gallery extending from the tip of the leaf and generally occupying its full width. Larva yellow; head pale brown.
P. 5-6. On detritus or the lower part of a leaf of the foodplant.
I. 6-7.

564(869). **E. pulchella** (Haw.) (*obscurella* Staint.)
O. On *Holcus mollis* or *Arrhenatherum elatius*; probably also on other Gramineae.
L. 3-5;7. In a flat mine. Larva pale yellowish grey; head light brown.
P. 4-5;7-8.
I. 5;8.

565(885). **E. rufocinerea** (Haw.)
O. On *Holcus mollis* or *Arrhenatherum elatius*.
L. 12-4. In a gallery directed either towards or away from the tip of the leaf, of variable width, but always flat. Larva dull yellow, dorsally greenish grey; head and narrowly divided prothoracic plate brown.
P. 3-4. On a leaf under a slight cocoon.
I. 4-5.

566(880). E. **cerusella** (Hübn.)
O. On *Phalaris arundinacea* or *Phragmites* (summer brood only).
L. 10-4;7. In a conspicuous, whitish blotch. Larva yellowish green; head and prothoracic plate brown.
P. 4-5;7-8. On a leaf or stem, under a girdle; often on the mined leaf, below the mine.
I. 5-6;8.

567(886). E. **argentella** (Clerck) (*cygnipennella* (Hübn.))
O. On one of various Gramineae, especially *Dactylis, Bromus sterilis* or *Phalaris arundinacea*.
L. 3-4. Mine varies according to the foodplant; in broad-leaved grasses, a short, transparent mine as broad as the blade; in narrow-leaved species such as *Festuca rubra*, the larva appears as a conspicuous swelling in the leaf. Larva dull greenish grey, dorsal line whitish; head and prothoracic plate, which consists of eight small sclerites in two plates, brown.
P. 4-5. Under a girdle on detritus.
I. 5-6. Flies from late afternoon till after dark, and comes to light.

568(884). E. **triatomea** (Haw.)
O. On *Festuca ovina* or *F. tenuifolia*.
L. 4-6. When young, in a gallery leading from the tip of the leaf; on changing mines, the larva generally mines upwards. The mine is at first yellowish grey with a violet reflection, later whitish and conspicuous. Larva yellowish grey, tinged olive-green; head pale brown, divided prothoracic plate paler than head.
P. 6.
I. 6-7.

569(882). E. **collitella** (Dup.)
O. On *Festuca ovina* or sometimes other grasses.
L. 6-7;10-5.
P.
I. 5-6;*7. Possibly univoltine in Britain.

570(881). E. **subocellea** (Steph.)
O. On *Brachypodium sylvaticum*.
L. 4-5. When young, in a *Stigmella*-like mine directed towards the tip of the leaf; later in a *Phyllonorycter*-like mine directed downwards. Larva dull green; head and divided prothoracic plate light brown.
P. 5-6.
I. 6-7.

571(–). E. **festucicolella** Zell. *Antenna* 2:22(1978)
O. On *Festuca* spp.
L.
P.
I. 5-7. Recently placed on the British list, but it now appears that supposed British specimens are not referable to this species.

572(883 partim). **E. triseriatella** Staint.
O. Possibly on *Festuca* spp.
L.
P.
I. 5-7.

573(883 partim). **E. dispunctella** (Dup.)
O. On *Festuca* spp.
L. 6-7;9-4.
P.
I. 4-6;7-9.

574(868). **E. bedellella** (Sirc.)
O. On *Helictotrichon pratense*.
L. 9-4;6-7. In an obscure mine, extending from the tip; the upper surface becomes slightly puckered and whitish, the lower surface being more or less tinged with purple. Larva at first yellowish, looking dark in the mine, later greyish green, but looking yellowish in the mine; head pale brown, prothoracic plate, which is enlarged posteriorly, darker.
P. 4-5;7-8. On a leaf-base or stem.
I. 5-6;8. On downs.

575(873 partim). **E. megerlella** (Hübn.)
O. Laid on the upperside of a leaf, often two or three close together, of *Brachypodium sylvaticum*; less often on *Bromus erectus* or other Gramineae.
L. 11-4;7. First in a narrow, puckered gallery directed upwards, with the frass packed at the base; later, frequently in a fresh leaf, in a broad gallery, directed upwards or, less often, downwards. Larva dull greenish grey; head and prothoracic plate brown.
P. 4-5;7-8. On a leaf or stem, under a girdle.
I. 5-6;8.

576(–). **E. cingillella** (Herr.-Schäff.) *Ent.Gaz.* **14**:159(1963)
O. *On *Milium effusum*.
L. *7-8;9-4. In a mine similar to that of *E. apicipunctella*.
P.
I. 5-7;*8-9. Little known in the British Isles and moth recorded only in June, so perhaps univoltine here.

577(871). **E. unifasciella** (Haw.) (*gangabella* sensu auctt., partim)
O. On *Dactylis*.
L. 10-5. In an opaque, *Phyllonorycter*-like mine, causing the tip of the leaf, where the frass is packed, to shrivel; basal leaves lying on the ground are preferred. Larva pale yellow; head and prothoracic plate light brown.
P. 4-5. Under a girdle on or in the ground.
I. 5-6.

578(870). **E. gangabella** Zell. (*taeniatella* Staint.)
O. On *Brachypodium sylvaticum*; * on *Dactylis*.
L. 9-4. In a very irregular gallery with a characteristic central tube of silk mixed with frass; hibernates full-fed from November in the mine. Larva olive-green; head and prothoracic plate dark brown.

P. 4. On a leaf, stem or detritus, under a girdle.
I. 5-6.

579(887). **E. subalbidella** Schläg.
O. On *Molinia caerulea*; possibly also on other grasses.
L. 8-4. In a long, narrow, flat, transparent gallery directed downwards, with a faint central tube of silk mixed with frass; hibernates full-fed in the mine. Larva dark olive-green; head and prothoracic plate almost black.
P. 4-5. On the ground, under a girdle.
I. 6-7.

580(873 partim). **E. revinctella** Zell. (*adscitella* Staint.)
O. On *Deschampsia cespitosa* or *Sesleria caerulea*.
L. 7;9-5. In a gallery directed either to or from the tip of the leaf, with the frass packed at the beginning of the mine; after hibernation, the larva mines a young leaf in a broad gallery, directed downwards from the tip. Larva dull yellowish green; head and prothoracic plate, which is strongly enlarged posteriorly, brown.
P. 5-6;7-8. On the ground.
I. 5-7;8.

581(872). **E. bisulcella** (Dup.) (*zonariella* Tengst.)
O. On *Deschampsia cespitosa*.
L. 4-6. In a gallery extending downwards from the tip of the leaf; at first it is narrow and slightly inflated, later broad and almost transparent; feeds at first very slowly and readily changes mines. Larva whitish grey; head brown, divided prothoracic plate darker.
P. 6. On the lower part of a leaf, under a girdle.
I. 7-8.

582(852). **Biselachista trapeziella** (Staint.)
O. On *Luzula pilosa* or *L. sylvatica*; possibly also other *Luzula* spp.
L. 9-5. In a broad, flat gallery; mines a fresh leaf after hibernation. Larva yellowish white with two dorso-lateral red lines.
P. 5-6. On a leaf.
I. 6-7.

583(853). **B. cinereopunctella** (Haw.)
O. On *Carex flacca*.
L. 10-4. In a long gallery directed downwards, usually filling the space between the midrib and the leaf margin. Larva whitish with red dorso-lateral spots. Will not change leaves.
P. 4-5. In the angle of a leaf, under a girdle.
I. 5-6.

584(874). **B. serricornis** (Staint.)
O. On *Carex sylvatica* or probably other *Carex* spp.
L. 9-5. Overwinters in a long red-brown mine; changes to a new leaf in the spring, when it mines towards the tip of the leaf in a series of short, parallel mines between ribs. Larva grey-green; head and prothoracic plate light brown.
P. 5-6.
I. 6-7.

585(879). **B. scirpi** (Staint.)
O. On *Scirpus maritimus* or *Juncus compressus*.
L. 4-5. In a short, broad, greenish white mine, with the frass at first packed at the start of the mine, but later in a thick, central line. Larva pale greenish yellow; head pale yellow.
P. 5-6. On the midrib of a leaf, seldom that mined, under a slight web.
I. 6-7.

586(877). **B. eleochariella** (Staint.)
O. On *Carex, Eleocharis* or *Eriophorum* spp. *Carex flacca* may be the principal foodplant in Britain.
L. 5.
P. 5-6.
I. 6-7.

587(875). **B. utonella** (Frey) (*paludum* (Frey))
O. On *Carex* spp., especially *C. acutiformis* or *C. paniculata*.
L. 3-5. In a short mine occupying half the width of the blade and generally situated halfway between the tip and the base. Larva yellowish green; head black, prothoracic plate light brown.
P. 5-6. In the angle of a leaf, sometimes a dead leaf, under a girdle.
I. 5-7. May be obtained by day by sweeping the foodplant.

588(878). **B. albidella** (Nyl.) (*rhynchosporella* (Staint.))
O. On *Eriophorum angustifolium* or *Carex* spp., especially *C. acuta*.
L. 4-5. In a gallery extending from the tip of the leaf, which is at first whitish, but later turns yellowish grey and more transparent. Larva pale green; head shining greyish black, prothoracic plate greyish brown. Sometimes found with the previous species, but appears much paler in the mine and is not full-grown till a fortnight later.
P. 5-6. On a leaf or stem, under a girdle.
I. 6-8.

589(867). **Cosmiotes freyerella** (Hübn.) (*nigrella* (Hübn.) nec (Fabr.))
O. On *Poa* spp.; probably also on other Gramineae.
L. 3-4;7. In a long, rather broad mine, extending from the tip of the leaf. Larva yellowish grey; head and prothoracic plate black.
P. 4-5;7-8. On a leaf or stem of the foodplant, under a dense web.
I. 4-6;8.

590(867A). **C. consortella** (Staint.) (*exiguella* (Frey)) *Ent.Gaz.* **3**:189-190 (1952)
O.
L.
P.
I. 3-9. In two, or perhaps three, generations.

591(866). **C. stabilella** (Staint.)
O. On *Agrostis* or, probably, other grasses.
L. 2-5. In a long, narrow, yellowish gallery, extending from the tip of the leaf, often with more than one larva in the same leaf.

Larva yellowish; head and prothoracic plate brownish.
P. 5-6. On a leaf or on the ground, under a dense web.
I. 5-7. Possibly bivoltine.

OECOPHORIDAE

592(751). **Schiffermuelleria grandis** (Desv.)
O.
L. 12-3. On decayed wood under bark, chiefly *Fagus* and *Quercus*.
P. 4-5. In the feeding place.
I. 5-6. Flies in sunshine between 1000 and 1300 hrs.

593(762). **S. subaquilea** (Staint.)
O.
L.
P.
I. 5-6. On heaths and mountains; flies in afternoon sunshine.

594(761). **S. similella** (Hübn.)
O.
L. 9-4. Under dead bark of *Pinus* or *Acer pseudoplatanus*; larva grey, plates brownish.
P. 5.
I. 6-7.

595(764). **S. tinctella** (Hübn.)
O.
L. 9-3. On dead wood and under decaying bark; on moss and lichen on trees.
P. 4. In the feeding place.
I. 5-6.

596(754). **Chambersia augustella** (Hubn.)
O.
L. 9-4. Under dead bark of *Ulmus, Malus, Tilia* or *Acer pseudoplatanus*; larva light ochreous grey; head dark brown; prothoracic plate lighter brown.
P. 5.
I. 6. Rests on trunks.

597(–). **Bisigna procerella** ([D. & S.]) *Ent. Rec.* 88:211-212(1976)
O.
L. 9-5. *On lichens on trunks of *Acacia, Fagus* etc.; larva olive-brown, head black.
P.
I. 7. Comes to light.

598(756). **Batia lunaris** (Haw.)
O.
L. 10-4. Under dead bark of *Malus, Crataegus, Robinia* etc.; in dead wood on fencing etc.; *also on mosses and lichens.
P. 5-6. In the feeding place.
I. 6-7. Comes to light.

599(757). **B. lambdella** (Don.)
O.
L. 9-5. In dead wood of *Ulex*; *also of *Carpinus, Aesculus, Alnus*. Larva grey-whitish.
P. 5-6.
I. 6-7. Rests after dark on dead stems of gorse, and comes to light.

600(756). **B. unitella** (Hübn.)
O.
L. 10-5. In decaying bark of *Ulmus, Quercus, Malus, Robinia* or in dead sticks; *also in lichens and mosses on trees. Larva dull dark brown, with whitish lines, head red-brown.
P. 6. In the feeding place.
I. 7-8. Has been taken at light.

601(753). **Dafa formosella** ([D. & S.])
O.
L. 9-4. On decayed wood, under bark, chiefly of *Malus*; larva light grey, head and plate light chestnut.
P. 5-6.
I. 7. Rests on fences etc. Probably extinct in Britain.

602(760). **Borkhausenia fuscescens** (Haw.)
O.
L. 10-4. On decaying leaves and in birds' nests.
P. 5-6.
I. 7-8.

603(763). **B. minutella** (Linn.)
O.
L. 9-3. On seeds and dry vegetable matter.
P. 4-5.
I. 5-6. Sometimes plentiful in reed-stacks or old haystacks.

604(755). **Telechrysis tripuncta** (Haw.)
O.
L. Larva unknown, but probably in rotten wood in hedges and thickets.
P.
I. 5-6. Flies in sunshine in woods.

605(768). **Hofmannophila pseudopretella** (Staint.)
O.
L. 6-4. On seeds, dried plants, wool, skins, dead insects, frass, books etc., in birds' nests; larvae are killed by severe frost.
P. 3-8. In the feeding place, in a cocoon of silk mixed with the material on which the larva has fed.
I. 5-9. A common household pest.

606(758). **Endrosis sarcitrella** (Linn.) (*lactella* ([D. & S.]))
O.
L. 1-12. In a silken gallery mixed with frass and food-particles, on dry vegetable matter, paper, rotten wood, fungus, dead insects, in birds' nests, in bee-hives, etc.
P. In the feeding place.
I. 1-12. More or less continuously brooded in houses. Out of doors, the adults sometimes overwinter in haystacks.

607(749). **Esperia sulphurella** (Fabr.)
O.
L. 7-3. On dead wood, exuding much frass; larva greyish white, spots blackish, head and plate reddish brown.
P. 3-5. In a cocoon of silk and frass in the feeding place or under bark.
I. 5-6. Males fly in morning sunshine, sometimes assembling to females.

608(750). **E. oliviella** (Fabr.)
O.
L. 8-4. On decayed wood of *Quercus* spp., *Prunus* spp., *Corylus, Robinia* etc.; also in rotten cut wood; larva pale yellowish grey with black dots; head and plate brown.
P. 5-6. In the feeding place.
I. 6-7. Flies in afternoon sunshine.

609(748). **Oecophora bractella** (Linn.)
O.
L. 2-4, but probably hatches in summer. On rotten wood of stumps and under the bark of various trees; possibly also on vegetable detritus; larva brownish grey, head pale brown.
P. 5.
I. 5-6.

610(747). **Alabonia geoffrella** (Linn.)
O.
L. 2-4. On decayed wood; larva whitish with darker spots, head and prothoracic plate yellow-brown.
P. 4-5.
I. 5-6.

611(773). **Aplota palpella** (Haw.)
O.
L. 10-6. Gregarious in silken tubes in moss growing on trunks of *Ulmus, Quercus* spp. or fruit trees; also on the moss *Homalothecium sericeum* growing on sandstone walls.
P. 6-7.
I. 7-8.

612(774). **Pleurota bicostella** (Clerck)
O.
L. 9-5. On shoots of *Erica*, in a web.
P. 5-6.
I. 6-7. Readily disturbed by day; flies in the evening.

613(–). **P. aristella** (Linn.) *Proc.Brit.ent.nat.Hist.Soc.* **1970**:37
Described as an uncertain British species by Stephens in 1829; there is no authentic British specimen.

614(772). **Parocystola acroxantha** (Meyr.)
O. 8-9.
L. 9-?. In captivity, in a web on the underside of fresh or withered leaves of *Eucalyptus*.
P.
I. 6-9. Comes to light. Has been gound on heathland, in gardens and in a flour-mill; may be bivoltine.

615(776). **Hypercallia citrinalis** (Scop.) (*christiernana* (Linn.))
O.
L. 5. In spun shoots of *Polygala vulgaris* and *P. calcarea*.
P. 5-6. Projects naked from a silken pad spun on a part of the foodplant.
I. 6-7.

616(775). **Carcina quercana** (Fabr.)
O. On *Quercus* spp., *Fagus, Malus, Castanea, Acer pseudoplatanus*, etc.
L. 5-6. Beneath a white silken web on the underside of a leaf. Larva bright green; subdorsal lines yellow-green.
P. 6-7. On the underside of a leaf, under a silk web.
I. 7-9. Comes to light.

617(759). **Amphisbatis incongruella** (Staint.)
O. On *Calluna*.
L. 6-7. In a case resembling a blackish section of grass-stem.
P. 8-4. In the case, attached to stones, tree-trunks, fences, etc.
I. 4. Flies only in sunshine on still days for about two hours before noon.

618(767 partim). **Pseudatemelia josephinae** *(Tubuliferola josephinae)* (Toll) *Bull.amat.ent.Soc.* **25**:88(1966); *Dt.ent.Z.* **6**:183-184(1959)
O.
L. Life history not yet differentiated from that of the next species.
P.
I. Late 6-early 8. The date of capture will generally distinguish this and the next species.

619(767 partim). **P. flavifrontella** ([D. & S.])
O. 6-7. Laid in clusters on decaying leaves; hatch in about 21 days.
L. 7-4. On decaying leaves and on lichens in a portable case made from a leaf fragment; lives two years. Larva yellowish white.
P. 5-6. In the case, often affixed to a tree-trunk.
I. End 5-6. Comes to light.

620(766). **P. subochreella** (Doubl.) (*panzerella* sensu Steph.)
O. 7. Laid in clusters on decaying leaves; hatch in about 21 days.
L. 7-4. On decaying leaves on the ground, moss, lichen and dead insects, in a portable case made of a leaf-fragment. To find the case, rake over the leaves in February or March in the angles at the base of oak-trees; the cases will be found lying on the ground. Larva reddish violet.
P. 5-6. In the case, spun on a dead leaf or on a trunk about 25 mm above ground level.
I. 6-7. The males fly freely at about noon.

Larvae of the next three species may be recognised by the legs of the third thoracic segment, which appear inflated and as if they were wearing boxing-gloves.

621(770). **Diurnea fagella** ([D. & S.])
O. On *Quercus* spp., *Fagus, Betula* spp., *Salix*, etc.
L. 6-10. In a loose spinning or a folded leaf; larva grey-green, head yellowish grey.
P. 11-3. In detritus.
I. 4. Males rest by day on trunks; females may sometimes be found low down on trunks at nightfall.

622(771). **D. phryganella** (Hübn.)
O. On *Quercus* spp.; also on *Vaccinium myrtillus*.
L. 6-7. In a folded or rolled leaf. Larva yellowish green, head brown.
P. 8-9. In detritus.
I. 10-11. Males fly by day in open woodland; females rest on trunks.

623(769). **Cheimophila salicella** (Hübn.)
O. On *Myrica, Potentilla anserina, Salix caprea, Prunus spinosa, Filipendula ulmaria*, etc.
L. 6-9. In a folded leaf. Larva whitish green, spots grey; head blackish.
P. 9-3. In detritus.
I. 4. Males fly in sunshine in the middle of the day and assemble to females.

624(822). **Semioscopis avellanella** (Hübn.)
O. On *Tilia cordata, Betula* spp. or *Carpinus*.
L. 7-9. On lime in a web beneath a leaf; on hornbeam rolling a leaf into an open-ended tube.
P.
I. 3-4. Comes to light.

625(821). **S. steinkellneriana** ([D. & S.])
O. On *Sorbus aucparia, Crataegus, Fraxinus* or *Prunus spinosa*.
L. 7-9. In a web beneath a leaf.
P.
I. 4. Flies freely at dawn.

626(823). **Enicostoma lobella** ([D. & S.])
O. On *Prunus spinosa* or, occasionally, on other rosaceous trees.
L. 8-9. In a web beneath the leaf, causing the edges to turn downwards; drops rapidly from its spinning if alarmed. Several leaves are attacked. Larva green; dorsal line darker, pale-edged; prothoracic plate with two black dots.
P. In earth.
I. 6.

627(778). **Depressaria discipunctella** Herr.-Schäff.
O. On *Heracleum sphondylium, Pastinaca* or *Angelica sylvestris*.
L. 6-7. In the flowers and fruits.
P.
I. 9-4.

628(779). **D. daucella** ([D. & S.]) (*nervosa* sensu auctt.)
O. On *Oenanthe crocata, O. aquatica* and, probably, other umbellifers.
L. 6-7. In spun flowers, later boring into the stem. Larva dark bluish grey.
P. 7-8. In the stem.
I. 9-4.

629(791). **D. ultimella** Staint.
O. On *Oenanthe aquatica, O. crocata* or *Sium*.
L. 6-8. In an umbel, soon passing into the stem. Larva whitish green, head black.
P. 7-9.
I. 9-4.

630(782). **D. pastinacella** (Dup.) (*heracliana* sensu auctt.)
O. On *Heracleum sphondylium, Pastinaca* or, occasionally, *Oenanthe silaifolia*.
L. 6-7. On the flowers and seeds, in a silken tube. Larva dark bluish grey; subdorsal line yellow.
P. In the stem.
I. 9-4.

631(787). **D. pimpinellae** Zell.
O. On *Pimpinella saxifraga* or *P. major*.
L. 7-8. On the flowers, in a slight spinning.
P. 8-9. In a cocoon in earth.
I. 9-4.

632(786). **D. badiella** (Hübn.)
O. On *Hypochoeris radiata, Sonchus arvensis* or *Taraxacum*.
L. 5-7. When young, feeding between spun leaves causing conspicuous transparent brownish blotches on the upper surface; later, in the rosette leaves from a gallery on the surface of the soil; finally, eating out heart of the plant which it frequently kills.
P. 7-8. In a cocoon in earth.
I. 8-9. Fond of resting on gorse or juniper.

633(819). **D. brunneella** Rag.
One specimen taken at New Romney, Kent in August, 1916.

634(784). **D. pulçherrimella** Staint.
O. On *Conopodium*.
L. 5-6. On the flowers and seeds; also stated to feed in rolled leaves of *Pimpinella*.
P. 6-7. In a cocoon in earth.
I. 7-8. Fond of resting on gorse or juniper.

635(785). **D. douglasella** Staint.
O. On *Daucus* or *Pastinaca.*
L. 5-6. In a leaf spun into a tube. Larva green, head yellow-green.
P. 6-7. In a cocoon in earth.
I. 7-8.

636(781). D. **weirella** Staint.
O. On *Conium, Anthriscus* or *Chaerophyllum*.
L. 5-6. In the leaves, folding the tips. Larva green, head green.
P. 6-7. In a cocoon in earth.
I. 7-8.

637(783). D. **emeritella** Staint.
O. On *Tanacetum vulgare*.
L. 6-7. In rolled leaves.
P. 7-8.
I. 8-9. "A specimen of this insect has occurred in England, but its locality is unknown." *Nat.Hist.Tin.* **6**: 230-239 (1861).

638(790). D. **albipunctella** (Hübn.)
O. On *Chaerophyllum* or *Anthriscus*. *Also on *Daucus, Conium* or *Pimpinella*.
L. 5-7. In rolled leaves. Larve green with grey spots; head and prothoracic plate green, spotted black.
P. 7-8. In a cocoon in earth.
I. 8-4.

639(789). D. **olerella** Zell.
O. On *Achillea millefolium*.
L. 6-7. On spun leaves. Larva dull green, stripes inconspicuous; head black.
P. 7-8.
I. 8-4.

640(780). D. **chaerophylli** Zell.
O. On *Chaerophyllum*.
L. 6-7. In a light web amongst or below the flowers, later passing into the stem.
P. 7-8. In a cocoon in earth.
I. 8-4.

641(788). D. **depressana** (Hübn.) (*depressella* Fabr.)
O. On *Daucus, Pastinaca* or *Pimpinella*.
L. 7-8. On the flowers.
P. 8-9.
I. 9-4 or later. Flies in afternoon.

642(–). D. **silesiaca** Hein. *Proc.Brit.ent.nat.Hist.Soc.* **1970**:37
O. On *Achillea millefolium*.
L. 7. Spins a tube about 30 mm long near the tip of a leaf, from which it feeds on the leaflets below. Larva bright green, stripes conspicuous; head mottled.
P. 7-8.
I. 8-4.

643(806). **Levipalpus hepatariella** (L. & Z.)
O.
L.
P.
I. 8.

644(816). **Exaeretia ciniflonella** (L. & Z.)
O. On *Betula* spp.
L. 4-5. In folded leaves.
P. 5-6.
I. 6-3. Rests on old stone walls.

645(777). **E. allisella** Staint. *Ent.* **95**:186(1962)
O. On *Artemisia vulgaris*.
L. 10-5. In the root and stem; in early spring it mines up a new shoot, causing it to wilt; changes readily to a fresh shoot.
P. 6. In debris at the base of the foodplant.
I. 7.

646(811). **Agonopterix heracliana** (Linn.) (*applana* (Fabr.))
O. On *Anthriscus, Heracleum sphondylium, Angelica sylvestris, Conium, Oenanthe* spp. or *Silaum*.
L. 5-7. In folded leaves. Larve green; head green; first thoracic segment with two lateral blackish green crescentic marks.
P. 7-8. In a cocoon in earth.
I. 8-4.

647(810). **A. ciliella** (Staint.)
O. On *Angelica sylvestris, Daucus, Heracleum sphondylium, Selinum* or *Silaum*.
L. 6-7. In spun or rolled leaves. Larva green, lines darker; head and plates brownish.
P. 7-8. In a cocoon in earth.
I. 8-4.

648(805). **A. cnicella** (Treits.)
O. On *Eryngium maritimum* or *E. campestre*.
L. 5-6. In spun shoots and rolled leaves, sometimes causing brown discolouration.
P. 6.
I. 6-7.

649(812). **A. purpurea** (Haw.)
O. On *Anthriscus, Daucus, Chaerophyllum* or *Torilis* spp.
L. 6-7. In rolled leaves. Larve yellowish; head and prothoracic plate black.
P. 7-8. In a cocoon in earth.
I. 8-5. Flies in sunshine after hibernation.

650(798). **A. subpropinquella** (Staint.)
O. On *Cirsium, Carduus, Centaurea* or *Arctium lappa*.
L. 6-7. In a web beneath a leaf; on knapweed, in a leaf rolled upwards. Larva green; head and prothoracic plate black; dots grey.
P. 7-8. In a cocoon in earth.
I. 8-4.

651(804). **A. putridella** ([D. & S.])
O. On *Peucedanum officinale*.
L. 5-6. In spun leaves, making conspicuous bunches, with a preference for young leaves.

P. 6-7. In spun leaves in the upper part of the plant.
I. 7-8. May be obtained by smoking the foodplant with a bee-smoker; flies freely in the evening.

652(801). **A. nanatella** (Staint.)
O. On *Carlina vulgaris*.
L. 4-6. At first, mining the leaves, making pale blotches; later, in a rolled leaf.
P. 6-7. In a cocoon in the earth.
I. 7-8.

653(813). **A. alstroemeriana** (Clerck)
O. On *Conium*.
L. 6-7. In rolled leaflets and in spun flowers.
P. 7-8. In a cocoon in earth.
I. 8-4.

654(800). **A. propinquella** (Treits.)
O. 4-5. On *Carduus*; *also on *Cirsium, Arctium* or *Serratula*.
L. 5-8. At first, mining a leaf, causing brown discolouration; later, in a web beneath a leaf. Larva green; head, prothoracic plate and pinacula black.
P. 8-9. In the feeding place or in a cocoon in earth.
I. 9-5.

655(799). **A. arenella** ([D. & S.])
O. On *Centaurea, Arctium* or *Carduus;* *also on *Carlina, Sonchus, Scabiosa* or *Knautia*.
L. 6-8. In a folded leaf. Larva green; head pale brown.
P. 8-9. In detritus on the ground.
I. 9-5.

656(794 partim). **A. liturella** ([D. & S.])
O. On *Centaurea nigra* or *Knautia*.
L. 5-6. In rolled leaves. Larva blackish.
P. 6-7. In a cocoon in earth.
I. 7-8.

657(794 partim). **A. bipunctosa** (Curt.)
O. On *Centaurea scabiosa* or *Serratula*.
L. 5-6. In a rolled leaf.
P. 6-7.
I. 7-8. Recently taken out of synonymy with the preceding species.

658(795). **A. pallorella** (Zell.)
O. On *Centaurea*; mainly, if not only, on *C. scabiosa*.
L. 6-7. In the rolled lobe of a leaf, generally the terminal lobe. Larva dull green; head red-brown.
P. 7-8.
I. 8-4.

659(814). **A. ocellana** (Fabr.)
O. On *Salix* spp.
L. 6-7. In spun or rolled leaves.
P. 7-8. In a cocoon in earth.
I. 9-4.

660(796). **A. assimilella** (Treits.)
O. On *Sarothamnus*.
L. 10-4. When young, in a short mine in a twig, in which it overwinters; in the spring between two twigs spun together at a joint. Larva brown or dark green; head and prothracic plate black.
P. 4-5. Between spun twigs.
I. 6-8.

661(797 partim). **A. pulverella** (Hübn.) (*atomella* (Hübn.) nec ([D. & S.]))
O. On *Genista tinctoria*.
L. 4-6. In leaves spun into a tube. Larva green; head pale brown.
P. 7-8.
I. 7-4.

662(797 partim). **A. scopariella** (Hein.) *Ent. Rec.* **91**:51-53(1979)
O. On *Sarothamnus*.
L. 6-7. In terminal shoots and spun leaves;Larva green; head pale brown.
P. 7-8.
I. 8-4. See reference for larval characters of nos 660, 661, 662 and 664.

663(793). **A. ulicetella** (Staint.) (*umbellana* sensu auctt.)
O. On *Ulex europaeus*.
L. 6-7. In a silken tube on young shoots. Larva greyish green; head black.
P. 7-8.
I. 8-4.

664(792). **A. nervosa** (Haw.) (*costosa* (Haw.))
O. On *Ulex*, *Sarothamnus* or *Genista*.
L. 5-6. In spun shoots. Larva green, yellowish or brownish yellow; head brownish or black; prothoracic plate greenish or yellowish brown.
P. 6-7. Sometimes on the foodplant; sometimes in a cocoon in earth.
I. 7-8.

665(794A). **A. prostratella** (Constant)
O. *On *Genista prostrata*, which is not recorded in Britain.
L. *Between twigs spun together.
P.
I. 8. One specimen taken in Ashdown Forest, Sussex in 1957.

666(802). **A. carduella** (Hübn.)
O. On *Carduus, Cirsium, Arctium* or *Centaurea nigra*.
L. 5-7. In a long, whitish mine near the midrib, from which frass is ejected; when on *Centaurea*, in a rolled leaf. Larva pale grey.
P. 7. In a cocoon in earth.
I. 7-8.

667(817). **A. liturosa** (Haw.) (*hypericella* sensu auctt.)
O. On *Hypericum perforatum, H. hirsutum* or *H. tetrapterum*.
L. 5-6. In spun shoots.
P. 6-7.
I. 7.

668(818). **A. conterminella** (Zell.)
O. On *Salix* spp.
L. 5-6. In a spun terminal shoot.
P. 6-7. In a cocoon in earth.
I. 7-9.

669(803). **A. curvipunctosa** (Haw.) (*zephyrella* (Hübn.))
O. On *Chaerophyllum* or *Anthriscus*.
L. 6-7. In spun flowers and leaves. Larva green, head yellowish.
P. 7-8.
I. 8-4.

670(820). **A. astrantiae** (Hein.) *Ent.Rec.***47**:57-58(1935)
O. On *Astrantia major* or *Sanicula europaeus*.
L. 5-6. In spun lower leaves.
P.
I. 7-8. Taken in Gloucestershire in 1933 and at Dungeness, Kent in 1961.

671(807). **A. angelicella** (Hübn.)
O. On *Angelica sylvestris* or *Heracleum sphondylium*.
L. 5-6. In spun shoots and leaves, making large, conspicuous bunches; often feeds gregariously.
P. 6-7.
I. 7-9.

672(815). **A. yeatiana** (Fabr.)
O. On *Daucus* or *Silaum;* *on *Carum* or *Oenanthe pimpinelloides.*
L. 6-7. On spun shoots or in rolled terminal leaves. Larva yellowish green, head black.
P. 8. In a slight cocoon on the surface of the soil.
I. 9-4.

673(809). **A. capreolella** (Zell.)
O. On *Pimpinella saxifraga*; *on *Daucus, Falcaria* or *Sium*.
L. 6. In spun radial leaves.
P. 7.
I. 8-4.

674(808). **A. rotundella** (Dougl.)
O. On *Daucus*.
L. 6-8. In rolled leaves. Larva green with darker dorsal and subdorsal lines, head black.
P. 8-9. In a cocoon in earth.
I. 9-4.

ETHMIIDAE

675(951). **Ethmia terminella** Fletch. (*sexpunctella* (Hübn.))
O. 7-8. On *Echium vulgare*, singly on any part of the plant; hatches in about 14 days.

L. 8-9. On the flowers and unripe seeds, in a silken gallery spun along the upperside of a flowering shoot, drawing the blossoms and sepals together so as to hide the gallery.
P. 10-5. In a dead stem, rotten wood or detritus.
I. 5-7. Rests by day on fences, telegraph poles, etc.

676(947). **E. dodecea** (Haw.) (*decemguttella* (Hübn.))
O. 8. On *Lithospermum officinale*.
L. 8-10. Singly or gregariously on the leaves, under a slight web.
P. 10-5. In white cocoon spun amongst detritus.
I. 5-8. Rests on the foodplant.

677(948). **E. funerella** (Fabr.)
O. 7-8. On *Symphytum officinale, S. tuberosum, Pulmonaria officinalis* or *Lithospermum officinale*.
L. 8-10. On the underside of a leaf, under a slight web.
P. 10-5. In a white cocoon spun amongst detritus.
I. 5-7. Rests on the foodplant. Comes to light.

678(949). **E. bipunctella** (Fabr.)
O. On *Echium vulgare, Symphytum* or other Boraginaceae, usually on a leaf. Hatches in about nine days.
L. 6-7;9. On the flowers and leaves, under a slight web.
P. 7-8;9-5. In a dead stem or rotten wood; most August pupae overwinter until the following summer.
I. 5-6;8-9. Rests on tree-trunks, fences, etc.; comes to light.

679(950). **E. pusiella** (Linn.)
O. *On *Lithospermum* or *Pulmonaria*, possibly overwintering in this stage.
L. *5-6. On leaves and flowers.
P. *6. In a white cocoon.
I. 6-8. Rests on the foodplant. Perhaps now extinct in Britain.

680(946). **E. pyrausta** (Pall.)
O. *6. On *Thalictrum aquilegifolium*, which occurs in Britain only as a garden introduction.
L. *6-7. Singly, in a slight web.
P. *8-5. In a whitish cocoon above ground.
I. 5-6. Diurnal; the male flies in sunshine. Perhaps now extinct in Britain.

GELECHIIDAE

681(574). **Metzneria littorella** (Dougl.)
O. On *Plantago coronopus*.
L. 9-3. On the seeds.
P. 4-5. In the stem of the seed-head.
I. 5-6.

682(575). **M. lappella** (Linn.)
O. On *Arctium lappa*.
L. 9-4. On the seeds.

P. 5. In a cocoon in the seed-head.
I. 6-7. Comes to light.

683(576). **M. aestivella** Zell. (*carlinella* (Staint.))
O. On *Carlina*.
L. 10-4. On the seeds. A few raised florets in the centre of a flower indicate the presence of a larva.
P. 5. In a cocoon in the seed-head.
I. 6-7.

684(577). **M. metzneriella** (Staint.)
O. On *Centaurea nigra*.
L. 10-4. On the seeds.
P. 5. In a cocoon in the seed-head.
I. 6-8. Comes to light.

685(578). **M. neuropterella** (Zell.)
O. On *Centaurea nigra* or *C. scabiosa*. Reputed to feed also on the seeds of *Cirsium acaule*.
L. 10-5. On the seeds.
P. 6. In a cocoon in the seed-head.
I. 7-8. Comes to light.

686(580;603). **Paltodora cytisella** (Curt.) (*clinosema* Meyr.)
O. On *Pteridium*.
L. 5-6. In a stem, making a slight swelling; often in a side-shoot, aborting it.
P. 6-7. In the feeding place.
I. 7. Comes to light.

687(579). **Isophrictis striatella** ([D. & S.]) (*tanacetella* (Schrank))
O. On *Tanacetum vulgare* or *Achillea ptarmica*.
L. 10-5. In the seed-head.
P. 6-7. In a dead stem.
I. 7-8.

688(584). **Apodia bifractella** (Dup.)
O. On *Pulicaria dysenterica* or *Inula conyza*.
L. 10-4. On the seeds.
P. 5-6. In the seed-head.
I. 7-8. Rests on the flowers of the foodplant in the evening.

689(586). **Eulamprotes atrella** ([D. & S.])
O. On *Hypericum perforatum*.
L. 5. In the stem.
P. 6. In a flat case made from a portion of the stem.
I. 7.

690(587). **E. unicolorella** (Dup.)
O.
L.
P.
I. 6-7.

691(606). **E. wilkella** (Linn.) (*pictella* (Zell.))
O. On *Cerastium holosteoides*.
L. 5;7. In a silken tube just under the surface of the ground.

P. 5-6;7-8.
I. 6;8. Frequents sandy places.

692(–). **Argolamprotes micella** ([D. & S.]) *Proc.S.Lond.ent.nat.Hist.Soc.* **1965**(2):42
O. On *Rubus idaeus* and possibly other *Rubus* spp.
L. 5-6. Possibly on the pith of a shoot.
P.
I. 7.

693(588). **Monochroa tenebrella** (Hübn.)
O. On *Rumex acetosella*.
L. 9-4. In a rootstock.
P. 4-6. Possibly in the larval habitation.
I. 6-7. Best obtained by sweeping, especially round bushes.

694(597). **M. lucidella** (Steph.)
O. On *Eleocharis palustris*.
L. 5-6. In the stem.
P. 6.
I. 6-7.

695(593). **M. palustrella** (Dougl.)
O. On *Rumex crispus*.
L. 5-7. In a rootstock.
P.
I. 7-8. Comes to light.

696(600). **M. tetragonella** (Staint.)
O. On *Glaux maritima*.
L. 4-5. In the stem, killing the plant.
P. 5-6.
I. 6-7. In salt-marshes.

697(590,591). **M. conspersella** (Herr.-Schäff.) (*morosa* Mühl.; *quaestionella* (Herr.-Schäff.))
O. On *Lysimachia vulgaris*.
L. 5. In spun shoots.
P.
I. 7-8. Comes to light.

698(–). **M. hornigi** (Staud.) *Proc.S.Lond.ent.nat.Hist.Soc.* **1963**:59;*Ent. Rec.* **99**:105(1977)
O. On *Polygonum* spp.
L. In the stem or side-branches.
P. In a white silken cocoon on the stem.
I. 6. Comes to light.

699(595). **M. suffusella** (Dougl.)
O.
L.
P.
I. 6-7. In fens.

700(599). **M. lutulentella** (Zell.) *Svensk Insektfauna* **10**:89(1928)
O. *On *Filipendula ulmaria*.

L. In the rootstock.
P.
I. 7. Comes to light.

701(596,598,604). **M. elongella** (Hein.) (*servella* sensu Meyr.; *micrometra* (Meyr.)) *Ent.* **68**:121-122(1935)
O. On *Potentilla anserina.*
L.
P.
I. 6.

702(589). **M. arundinetella** (Staint.)
O. On *Carex acutiformis.*
L. 3-5. In a long gallery in a leaf.
P. 5-6. In a cocoon on a leaf of the foodplant just above water-level.
I. 6-7.

703(594). **M. divisella** (Dougl.)
O. *On *Iris* spp.
L.
P.
I. 6-7.

704(607). **Chrysoesthia drurella** (Fabr.) (*hermannella* sensu auctt.)
O. On *Chenopodium* and *Atriplex* spp., on the upperside of a leaf.
L. 7;9. Mining the leaves, in a gallery so contorted as sometimes to form a blotch; frass greenish and remaining in the mine.
P. 6;8. In detritus at ground level.
I. 6;8.

705(608). **C. sexguttella** (Thunb.) (*stipella* sensu Hübn.)
O. On *Chenopodium* and *Atriplex* spp.
L. 6;9-10. Mining the leaves, making white blotches from which the frass has generally been ejected.
P. 5-6;7. In detritus at ground level.
I. 5-6;8.

706(585). **Ptocheuusa paupella** (Zell.) (*inopella* sensu auctt.)
O. On *Pulicaria dysenterica, Inula crithmoides, Centaurea nigra* or *Mentha* spp.
L. 7;9-10. On the seeds. A few raised florets in the centre of a blossom of *P. dysenterica* indicate a larva.
P.
I. 6;8-9.

707(581). **Sitotroga cerealella** (Ol.)
O. Laid singly or in clusters on the sides of grains; hatching in 4 days to 4 weeks.
L. 10-3. On stored grains of maize, rice and wheat.
P. 3-5. In a white silk cocoon in the feeding place.
I. 5-6. A pest species imported with the grain.

708(582). **Psamathocrita osseella** (Staint.)
O.
L.
P.
I. 6-7. Associated with *Origanum*.

709(583). **P. argentella** P. & M. *Ent.Gaz.* **25**:302-304(1974)
O.
L. Probably on *Agropyron pungens*.
P.
I. 6-7. Flies before sunrise between 0445 and 0545 hrs and again at dusk; rests on the stems of the supposed foodplant.

710(601). **Aristotelia subdecurtella** (Staint.)
O. On *Lythrum salicaria*.
L. 6. In spun shoots.
P. 6-7.
I. 7.

711(602). **A. ericinella** (Zell.)
O. On *Calluna*.
L. 6-7. In a silken tube amongst shoots of the foodplant.
P. 7.
I. 7-8. Comes to light.

712(605). **A. brizella** (Treits.)
O. On *Armeria maritima* or *Limonium vulgare*.
L. 6-7;9-10. In the flower-head.
P. 4-5;7. In the flower-head.
I. 5-6;8.

713(592). **Xystophora pulveratella** (Herr.-Schäff.)
O. On *Lotus* spp., *Coronilla varia* or *Medicago* spp.
L. 8-9. Between spun leaves.
P.
I. 5-6.

714(609). **Stenolechia gemmella** (Linn.)
O. On *Quercus* spp.
L. 6. In a shoot.
P. 7.
I. 8-9. Rests in the daytime in a crevice in bark.

715(610). **Parachronistis albiceps** (Zell.)
O. On *Corylus*.
L. 5. In a bud.
P. 6.
I. 7-8.

716(611). **Recurvaria nanella** ([D. & S.])
O. On *Malus, Prunus spinosa* or cultivated species of *Prunus*.
L. 8-5. When young, mining a leaf; after hibernation, in a bud.
P. 6.
I. 7. Sometimes a pest in orchards. Rests on a trunk in the daytime; comes to light.

717(612). **R. leucatella** (Clerck.)
O. On *Crataegus* or *Malus*.
L. 5-6. Between two spun leaves.
P. 6-7.
I. 7. Comes to light.

718(612A). **Pulicalvaria piceaella** (Kearfott) *Ent.Rec.*73:40-41(1961)
O. On *Picea abies*.
L. 9-4. Mining loosely spun needles; no frass is left in the mine.
P.
I. 7. Comes to light.

719(613). **Exoteleia dodecella** (Linn.)
O. On *Pinus sylvestris*.
L. 9-5. First mining the apical half of a needle; the mine contains some spinning, has a hole at each end, and most of the frass is ejected. After hibernation, in shoots and spun needles.
P. 5-6.
I. 6-7. Comes to light.

720(614). **Rhynchopacha tetrapunctella** (Thunb.) (*lathyri* (Staint.))
O. On *Lathyrus palustris*.
L. 8-9. In a slight web.
P.
I. 6-7.

721(615). **R. mouffetella** (Linn.)
O. On *Lonicera* spp.
L. 5-6. In a white web spun along a twig and leaves.
P. 6-7.
I. 7.

722(616). **Xenolechia aethiops** (H. & W.)
O. On *Erica* spp.
L. 6-7. In a silken gallery amongst shoots.
P.
I. 5-6. Imagines congregate on burnt parts of heaths.

723(618). **Pseudotelphusa scalella** (Scop.)
O. On moss.
L. 10-4. On the moss.
P. 4-5.
I. 5-6. Rests on a trunk in the daytime.

724(623). **Teleiodes vulgella** (Hübn.)
O. On *Crataegus* or *Prunus spinosa*.
L. 4-5. Between spun leaves.
P. 5-6.
I. 6-7.

725(625). **T. scriptella** (Hübn.)
O. On *Acer campestre*.
L. 8-9. In a leaf folded upwards.
P. 10-5.
I. 6.

726(617). **T. decorella** (Haw.) (*humeralis* (Zell.))
O. On *Quercus* spp. or *Swida*.
L. 5-6. In a folded leaf.
P. 6-7.
I. 7-4.

727(619). **T. notatella** (Hübn.)
O. On *Salix caprea*.
L. 8-9. Eating the parenchyma under the down on the underside of a leaf, or between joined leaves.
P. 10-5.
I. 5-6.

728(–). **T. wagae** (Nowicki)
O.
L. Possibly on *Corylus*.
P.
I. 5. Well established in the Burren, Co. Clare. Formerly confused with *T. paripunctella*; a paper on this species is in preparation.

729(620). **T. proximella** (Hübn.)
O. On *Betula* spp. or *Alnus glutinosa*.
L. 8-9. In a folded leaf.
P. 10-5.
I. 5-6. Rests on a trunk in the daytime; comes to light.

730(621). **T. alburnella** (Zell.)
O. On *Betula* spp.
L. 5. In spun or folded leaves.
P. 6.
I. Late 6-9. Rests on trunks of birch in crevices of the bark, especially on rough bark.

731(622). **T. fugitivella** (Zell.)
O. On *Ulmus* spp.
L. 5. On the leaves.
P. 5-6.
I. 6-7. Rests on the trunks in crevices of the bark; comes to light.

732(624). **T. paripunctella** (Thunb.) (*triparella* (Zell.))
O. On *Quercus* spp. or *Myrica*.
L. 8-9. Between spun leaves, skeletonizing them.
P. 10-5.
I. 5-6.

733(626). **T. luculella** (Hübn.)
O. On *Quercus* spp.
L. 9. Between spun leaves.
P.
I. 5-6. Rests on the trunks in the daytime.

734(627). **T. sequax** (Haw.)
O. On *Helianthemum chamaecistus*.
L. 5-6. In spun shoots.
P. 6-7.
I. 7.

735(640). **Teleiopsis diffinis** (Haw.)
O. On *Rumex acetosella*.
L. 8-4;7-8. In a silken tube at the base of the stems.
P. 4-5;9.
I. 5-6;9-10. Comes to light.

736(628). **Bryotropha basaltinella** (Zell.)
O. On moss.
L. 3-4. On the moss.
P.
I. 6-7. Frequents old thatches.

737(630). **B. umbrosella** (Zell.)
O. On moss.
L. 5. On the moss.
P. 5-6.
I. 6-7. Comes to light.

738(631). **B. affinis** (Haw.)
O. On moss.
L. 1-4. On moss growing on walls.
P. 5-6. In the moss.
I. 6-7. Comes to light.

739(632). **B. similis** (Staint.)
O. On moss.
L. 4-5. On moss growing on walls.
P. 5-6.
I. 7.

740(633;634). **B. mundella** (Dougl.) (*portlandicella* Rich.)
O.
L.
P.
I. 5-6. On coast sandhills.

741(635). **B. senectella** (Zell.)
O. On moss.
L. 4-5. On moss growing on the ground.
P. 5-6.
I. 7.

742(655). **B. boreella** (Dougl.)
O.
L.
P.
I. 6-7.

743(656). **B. galbanella** (Zell.)
O. On moss.
L. 4-5. On the moss.
P. 6.
I. 7.

744(638). **B. figulella** (Staud.)
O.
L.

P.
I. 7. Status in Britain uncertain.

745(636). **B. desertella** (Dougl.)
O.
L.
P.
I. 6-7.

746(639). **B. terrella** ([D. & S.])
O. On Gramineae.
L. 3-4. In a silken gallery at the base of the stems.
P. 5-6.
I. 6-7. Comes to light.

747(637). **B. politella** (Staint.)
O.
L.
P.
I. 6. In dry places.

748(629). **B. domestica** (Haw.)
O. On moss.
L. 3-4. On moss growing on walls, in a silken gallery. Frass on top of the moss indicates a larva. On a moist morning the larva is often on the outside of the moss.
P. 5-6. In the moss.
I. 7-8. Rests on or near moss. Frequents old thatches.

749(647). **Chionodes fumatella** (Dougl.)
O. *On moss.
L. 5. *On the moss.
P.
I. 6-7. On sandy coasts. Comes to light.

750(648). **C. distinctella** (Zell.)
O. *On *Artemisia campestris*.
L. *In spun tubes in the sand under the foodplant.
P.
I. End 6-8.

751(641). **Mirificarma mulinella** (Zell.)
O. On *Ulex europaeus* or *Sarothamnus*.
L. 4-5. On the flowers; a small hole in a petal indicates a larva in the blossom.
P. 5-6.
I. 7-8. Comes to light.

752(659). **M. lentiginosella** (Zell.)
O. On *Genista tinctoria*.
L. 5-6. In spun shoots.
P. 6-7.
I. 7-8.

753(642). **Lita virgella** (Thunb.) (*longicornis* (Curt.))
O. On *Calluna*.
L. 4-5. On the leaves.
P. 5
I. 5-6.

754(657). **L. solutella** (Zell.)
O. On *Genista tinctoria*.
L. 7. In a silken tube.
P.
I. 5.

755(644). **Aroga velocella** (Zell.)
O. On *Rumex acetosella*.
L. 9-4;6. In a silken tube at the base of the stems.
P. 4-5;7.
I. 5;8. Flies freely in the afternoon; comes to light.

756(643). **Neofaculta ericetella** (Geyer)
O. On *Erica* spp. or *Calluna*.
L. 9-3. In spun shoots.
P. 4.
I. 5-6. Comes to light.

757(658 partim). **Neofriseria peliella** (Treits.) *Ent.mon.Mag.* **32**:248-249(1896)
O. On *Rumex acetosella*.
L. 5. In a silken tube at the base of the stems; larva dark red-brown.
P. 6.
I. 7.

758(658 partim). **N. singula** (Staud.) (*suppeliella* Wals.) *Ent.mon.Mag.* **32**:250 (1896)
O. On *Rumex acetosella*.
L. 5. In a silken tube at the base of the stems; larva green.
P. 6.
I. 7.

759(645). **Gelechia rhombella** ([D. & S.])
O. On *Malus*.
L. 5-6. Between joined leaves or in a folded leaf.
P. 6-7.
I. 7. Comes to light.

760(649). **G. scotinella** Herr.-Schäff.
O. On *Prunus spinosa*.
L. 4. In spun flowers.
P. 5.
I. 6-7. Comes to light.

761(–). **G. sabinella** Zell. *Ent.Gaz.* **29**:136-138(1978)
O. On *Juniperus*, in Britain probably on horticultural forms.
L.
P.
I. 8. Comes to light. One specimen was taken in North London in 1971 and another in Worcestershire in 1978.

762(650). **G. sororculella** (Hübn.)
O. On *Salix* spp.
L. 5-6. In spun shoots.
P. 6-7.
I. 7.

763(653). **G. muscosella** Zell.
O. On *Populus* and *Salix* spp.
L. 5. In a catkin.
P. 5-6.
I. 6-7.

764(652). **G. cuneatella** Dougl.
O. On *Salix* spp.
L. 7.
P. 8.
I. 9.

765(646). **G. hippophaella** (Schrank)
O. On *Hippophae rhamnoides*.
L. 6-7. In spun shoots.
P. 7-8.
I. 8-9.

766(654). **G. nigra** (Haw.)
O. On *Populus tremula* or *P. alba*.
L. 5-6. Between two spun leaves, one above the other.
P. 6. Between spun leaves.
I. 6-7. Rests on a trunk in the daytime.

767(651). **G. turpella** ([D. & S.]) (*pinguinella* (Treits.))
O. On *Populus nigra*, including var. *italica*.
L. 5. Between spun leaves.
P. 6.
I. 7. Rests on a trunk in the daytime but runs rapidly up the trunk or flies away when approached.

768(660). **Platyedra subcinerea** (Haw.) (*vilella* (Zell.))
O. On *Malva sylvestris*.
L. 6-7. On the flowers and seeds; a seed containing a larva has the sepals sealed down so as to cover the seed completely.
P. 7-8.
I. 8-5. Imagines fond of hiding in thatch.

769(661). **Pexicopia malvella** (Hübn.)
O. On *Althea officinalis* and cultivated hollyhocks.
L. 9-4. On the seeds; overwinters full-fed in its cocoon.
P. 5-6.
I. 6-7. Comes to light.

770(662). **Scrobipalpa suaedella** (Rich.)
O. On *Suaeda vera* or *S. maritima*.
L. 5. In a gallery amongst the leaves and blossom.
P. 6. In a cocoon on the ground.
I. 7.

771(663). **S. samadensis plantaginella** (Staint.)
O. On *Plantago coronopus* or *P. maritima*.
L. 4-5;7-8. In a root and in blotches in the leaves. Frass in the centre of a plant indicates a larva in the root.
P. 5-6;8-9. In the root.
I. 6-7;9.

772(668). **S. instabilella** (Dougl.)
O. On *Halimione*.
L. 3-5. In the leaves, making pale greenish blotches.
P. 5-6.
I. 6-7.

773(664). **S. salinella** (Zell.) (*salicorniae* (Her.))
O. On *Salicornia, Aster tripolium, Suaeda* or *Spergularia*.
L. 4-5. On *Aster* making a series of galleries in the leaves, similar to those of *Bucculatrix maritima*; larva with red dorsal line.
P. 6-8. Amongst detritus.
I. 6-9.

774(665). **S. ocellatella** (Boyd)
O. On *Beta vulgaris* subsp. *maritima*.
L. 1-5;7. On the buds, stems and in spun or mined leaves.
P. 5-6;7-8.
I; 6;8.

775(667). **S. nitentella** (Fuchs) (*seminella* P. & M.)
O. On *Chenopodium* spp. and *Suaeda maritima*.
L. 9-10. On the seeds, in a silken gallery.
P. 10-7. In a cocoon on the ground.
I. 8.

776(666). **S. obsoletella** (Fisch.v.Rösl.)
O. On *Atriplex* and *Chenopodium* spp.
L. 6-10. In the stem, eating the pith.
P. In the stem; a small hole indicates the presence of a pupa.
I. 5-9.

777(–). **S. clintoni** Povolny *Ent.Gaz.* **19**:113-118(1968)
O. On *Rumex crispus*.
L. 7. In the stem, ejecting quantities of pale brown frass.
P. 8-5. In the stem, where a small hole betrays its presence.
I. 5-6. Frequents sandy beaches.

778(669). **S. atriplicella** (Fisch.v.Rösl.)
O. On *Atriplex* and *Chenopodium* spp.
L. 5-6;10. In spun leaves and on the flowers and seeds.
P. 6-7;10-5.
I. 5;7-8.

779(670). **S. costella** (H. & W.)
O. On *Solanum dulcamara*, close to the midrib.
L. 8-9. In an opaque blotch containing frass, starting from the midrib; later between spun leaves, or in a berry or stem.
P. 9.
I. 9-6. Comes to light.

780(673). **S. artemisiella** (Treits.)
O. On *Thymus drucei*.
L. 5-6. In spun shoots.
P. 6.
I. 6-7.

781(–). **S. murinella** (Herr.-Schäff.) *Ent.Gaz.* **15**:79-82(1964)
O. On *Antennaria*.
L. 7-9. On the flower-heads and in blotch mines in leaves, moving from leaf to leaf; also in the stems.
P.
I. 5-6.

782(674). **S. acuminatella** (Sirc.)
O. On *Carduus, Cirsium* and *Centaurea* spp.
L. 7;9. On a lower leaf in a branched mine extending along the midrib; the mine is brownish.
P. 7-8;10-5. In litter on the ground.
I. 5-6;8. Comes to light.

783(–). **Scrobipalpula psilella** (Herr.-Schäff.) *Ent.Gaz.* **16**:9-10(1965)
O.
L. *On Compositae, including *Artemisia* spp., *Erigeron acer* and *Aster tripolium*.
P.
I. 6-7. Three specimens (1850, 1897 and 1926) have been taken in north Kent and south Essex.

784(672). **Gnorimoschema streliciella** (Herr.-Schäff.)
O.
L.
P.
I. 6. On heaths.

785(671). **Phthorimaea operculella** (Zell.)
O. On *Solanum tuberosum* (potato) and *Lycopersicon esculentum* (tomato).
L. 8-9. In the stems and leaves when young, later in the tubers; also on the fruits of tomato.
P. 9-5. In a cocoon on or near a tuber or fruit.
I. 6-7. Imported with the tubers and fruits.

786(686). **Caryocolum vicinella** (Dougl.) (*leucomelanella* sensu auctt.)
O. On *Silene maritima*.
L. 5. In spun shoots and leaves.
P. 6. In a cocoon on the ground, often attached to a stone.
I. 7-9.

787(683). **C. alsinella semidecandrella** (Threlfall)
O. On *Cerastium semidecandrum*.
L. 5. On the flowers and seeds.
P. 5-6.
I. 6-7. On coasts.

788(680). C. **viscariella** (Staint.)
O. On *Silene dioica, S. alba* or *Lychnis viscaria*.
L. 4-5. In a spun central shoot, the larva living in the central stem when not feeding.
P. 5-6. In detritus on the ground.
I. 6-7. Comes to light.

789(684). C. **marmoreum** (Haw.)
O. On *Cerastium holosteoides*.
L. 3-4. On leaves touching the ground, the larva living in a silken tube in the sand.
P. 5. In a cocoon in the sand.
I. 6-8. In sandy places, both coastal and inland.

790(679). C. **fraternella** (Dougl.)
O. On *Stellaria graminea, S. alsine* or *Cerastium holosteoides*.
L. 4-5. In spun shoots, making white blotches on the leaves.
P. 5-6. In detritus on the ground.
I. 6-7. Comes to light.

791(682). C. **proximum** (Haw.) (*maculiferella* (Dougl.))
O. On *Cerastium holosteoides*.
L. 5. On the flowers and seeds.
P. 6. In detritus on the ground.
I. 7.

792(677). C. **blandella** (Dougl.) (*maculea* sensu Haw.)
O. On *Stellaria holostea*.
L. 4-6. In spun shoots and seed-capsules. The capsules containing larvae are slate-coloured.
P. 6-7.
I. 7-8.

793(685). C. **junctella** (Dougl.) *Z.wien.ent.Ges.* **39**:287(1954)
O. *On *Stellaria* spp.
L.
P.
I. 8-4.

794(681). C. **tricolorella** (Haw.)
O. On *Stellaria holostea*.
L. 12-4. In spun shoots.
P. 5-6. In detritus in the ground.
I. 6-8.

795(678). C. **blandulella** (Tutt) *Opusc.ent.* **30**:17(1965)
O. *On *Cerastium pumilum*.
L.
P.
I. 7-8. On sandhills.

796(675). C. **kroesmanniella** (Herr.-Schäff.) (*huebneri* sensu auctt.)
O. On *Stellaria holostea* or *S. alsine*.
L. 5. In spun shoots.
P. 6.
I. 7-8.

797(676). **C. huebneri** (Haw.) (*knaggsiella* Staint.)
O. On *Stellaria holostea*.
L. 5. In spun shoots.
P. 6.
I. 7-8.

798(687). **Nothris verbascella** (Hübn.)
O. On *Verbascum pulverulentum*.
L. 9-7. In a web on the leaves, generally on the underside.
P. 6-7. In a cocoon on the underside of a leaf of the foodplant.
I. 7. Larvae, pupae and imagines may be found simultaneously on the foodplant in July.

799(–). **N. congressariella** (Braund) *Ent.Gaz.***9**:126-127(1958)
O. On *Scrophularia auriculata*.
L. 7-9;10-5. In spun leaves.
P.
I. 5-6;9-10.

800(688). **Reuttia subocellea** (Steph.)
O. On *Origanum* or *Mentha aquatica*.
L. 8-11. On the seed-heads, in a portable case made of florets or husks.
P. In the case.
I. 7-8.

801(689). **Sophronia semicostella** (Hübn.) *Opusc.ent.***30**:13(1965)
O. *On *Anthoxanthum odoratum*.
L.
P.
I. 6-7. Flies in the early afternoon and again at dusk.

802(690). **S. humerella** ([D. & S.])
O. On *Artemisia campestris*.
L. 5. Between spun leaves.
P. 6.
I. 7. Two specimens taken many years ago.

803(691). **Aproaerema anthyllidella** (Hübn.)
O. On *Anthyllis, Onobrychis* and *Trifolium* spp.
L. 4-5;7. On the leaves, making white blotches; in July, on the blossom and seed-pods.
P. 5;7-8. In May, in detritus on the ground; in July and August, in the flower-heads of *Anthyllis* and perhaps the other foodplants.
I. 5-6;8. Comes to light.

804(697A). **Syncopacma larseniella** (Gozmany) (*ligulella* sensu auctt.)
O. On *Lotus corniculatus, L. uliginosus* or *Genista* spp.
L. 5-6. In spun shoots.
P. 6. In detritus on the ground.
I. 6-7.

805(694,695). **S. sangiella** (Staint.) (*coronillella* sensu auctt., partim)
O. On *Lotus corniculatus*.
L. 5-6. Between joined leaves.

P.
I. 6-7.

806(692,694,695). **S. vinella** (Bankes) (*coronillella* sensu auctt., partim; *cincticulella* sensu auctt.)
O. On *Coronilla varia* or *Genista tinctoria*.
L. 5. Between spun leaves.
P.
I. 6-7.

807(698). **S. taeniolella** (Zell.)
O. On *Lotus corniculatus, Trifolium* or *Medicago* spp.
L. 5-6. In spun shoots and leaves.
P. 6. In detritus on the ground.
I. 7. Comes to light.

808(693). **S. albipalpella** (Herr.-Schäff.)
O. On *Genista anglica*.
L. 9-6. In spun leaves, making white blotches.
P. 6-7.
I. 7.

809(697). **S. cinctella** (Clerck) (*vorticella* Scop.)
O. On *Lotus corniculatus, L. uliginosus* or *Genista*.
L. 5-6. In spun shoots.
P. 6. In detritus on the ground.
I. 6-7. Comes to light.

810(–). **S. polychromella** (Reb.) *Ent.Gaz.*4:37(1953)
O.
L.
P.
I. A north African species which has been captured in Kent; probably imported accidentally.

811(700). **Acanthophila alacella** (Zell.)
O. On lichens.
L. 6. On lichens on tree-trunks.
P. 6-7.
I. 7-8. Comes to light.

812(699). **Anacampsis temerella** (L. & Z.)
O. On *Salix repens*.
L. 6. In spun shoots.
P. 6-7.
I. 7.

813(701). **A. populella** (Clerck)
O. On *Populus tremula* and *Salix* spp.
L. 5-6. In a rolled leaf.
P. 6-7. In the rolled leaf.
I. 7-8.

814(701A). **A. blattariella** (Hübn.) (*betulinella* Vari)
O. On *Betula* spp.
L. 5-6. In a rolled leaf.
P. 6-7. In the rolled leaf.
I. 7-9.

815(702). **Acompsia cinerella** (Clerck)
O. On moss.
L. 5-6. In moss at the base of a tree.
P. 6.
I. 6-7.

816(703). **Anarsia spartiella** (Schrank)
O. On *Ulex europaeus, Sarothamnus* or *Genista tinctoria*.
L. 5-6. In spun shoots.
P. 6-7.
I. 7. Comes to light.

817(–). **A. lineatella** Zell. *Ent.Gaz.* **10**:57(1959)
O.
L.
P.
I. A specimen was bred in 1957 from an imported apricot.

818(704). **Hypatima rhomboidella** (Linn.) (*conscriptella* (Hübn.))
O. On *Betula* spp. and *Corylus*.
L. 5-6. In spun shoots and rolled leaves.
P. 6-7. In detritus on the ground.
I. 7.

819(705). **Psoricoptera gibbosella** (Zell.)
O. On *Quercus* spp.
L. 5-6. On the leaves.
P. 6-7.
I. 7-8. Rests on a trunk in the daytime; comes to light.

820(706). **Mesophleps silacella** (Hübn.)
O. On *Helianthemum chamaecistus*.
L. 8. In a seed capsule.
P.
I. 6-7.

821(707). **Telephila schmidtiellus** (Heyd.)
O. On *Origanum*.
L. 6. In a rolled or folded leaf.
P. 6-7. Sometimes in a folded leaf.
I. 7-8.

822(708). **Dichomeris marginella** (Fabr.)
O. On *Juniperus*.
L. 5-6. In a web amongst the needles.
P. 6-7.
I. 7-8. Rests in juniper bushes.

823(709). **D. juniperella** (Linn.)
O. On *Juniperus*.
L. 4-5. In a web amongst the needles.
P. 6.
I. 7.

824(710). **D. ustalella** (Fabr.)
O. On *Carpinus, Betula* spp. or *Salix* spp.
L. 9-10. Between joined leaves.

P.
I. 5-6.

825(711). **D. fasciella** (Hübn.)
O. On *Prunus spinosa*.
L. 9-10. In a leaf folded upwards.
P.
I. 5-6.

826(714). **Brachmia blandella** (Fabr.) (*gerronella* Zell.) *Ent.Rec.* **90**:67-68 (1978)
O. On *Ulex europaeus* and probably other plants.
L. 6. On the leaves; as an inquiline in a gall on *Abies grandis*.
P. 6-7.
I. 7. Comes to light.

827(715). **B. inornatella** (Dougl.)
O. On *Phragmites*.
L. 9-4. In a stem.
P. 5.
I. 6. In fens.

828(716). **B. rufescens** (Haw.)
O. On *Poa* spp., *Arrhenatherum* and many other grasses.
L. 8-5. In a rolled leaf, blanching the tip; the larva moves from leaf to leaf.
P. 6. In a rolled leaf, not the one on which the larva has fed.
I. 7. Comes to light.

829(717). **B. lutatella** (Herr.-Schäff.)
O. On *Calamagrostis epigejos*.
L. 5-6. In rolled leaves.
P. 6-7.
I. 7-8.

830(712 partim). **Oegoconia quadripuncta** (Haw.)
O.
L. 5-6. On decaying vegetable matter.
P. 6-7. Amongst the food material.
I. 7-8. Sometimes plentiful in hedge-bottoms.

831(712 partim). **O. deauratella** (Herr.-Schäff.) *Ent.Rec.* 78:243(1966)
O.
L. Habits similar to those of the previous species.
P.
I.

832(713). **Symmoca signatella** (Herr.-Schäff.)
O.
L. 5-6. On dry vegetable matter.
P. 6-7.
I. 7-9. An accidentally introduced species, perhaps now established, occurring near the London docks.

BLASTOBASIDAE

833(745). **Blastobasis lignea** Wals.
O. 8-9. Hatches in 14 days.
L. 9-6. On *Taxus, Picea, Abies, Juniperus* etc., on fresh and decaying leaves; in bird-droppings; probably on many other pabula.
P. 6-7. In a cocoon amongst the leaves of the foodplant or amongst litter on the ground.
I. 8. Comes to light.

834(745A). **B. decolorella** (Wollaston)
O. 6-7;10. Laid in small batches on detritus, hatching in about seven days.
L. 6-8;10-5. On fallen and decaying leaves, dead insects etc., in a thick silken tube; on the seeds inside the pods of *Lupinus arboreus*; on stored apples.
P. 6;8-9. In a thick silken cocoon amongst litter or in the feeding place.
I. 6-7;9-11. Comes to light.

835(745B). **B. phycidella** (Zell.) *Ent.Rec.* **61**:113(1949)
I. 6. Four specimens taken at Southampton in 1930 were probably imported accidentally.

836(746). **Auximobasis normalis** Meyr.
O.
L. Possibly in stored products.
P.
I. 9. A specimen taken in the Liverpool docks in 1921 was probably imported accidentally.

STATHMOPODIDAE

837(831). **Stathmopoda pedella** (Linn.)
O. 7-8.
L. 9. In fruits of *Alnus glutinosa*, sometimes causing slight discolouration and the extrusion of frass.
P. 10-6. In a cocoon in detritus on the ground.
I. 7. Can be obtained by tapping the branches of alder over a beating tray.

MOMPHIDAE

838(743). **Batrachedra praeangusta** (Haw.)
O. On *Populus* and *Salix* spp.
L. 4-6. In the catkins.
P. 6-7. In a cocoon, sometimes spun on the trunk.
I. 7. Rests on the trunks.

839(744). **B. pinicolella** (Zell.)
O. On *Picea abies* or *Pinus sylvestris*.
L. 9-5. On the needles, in a silken gallery on the surface of a twig.
P. 6. In a cocoon on the underside of a twig.
I. 7.

840(723). **Mompha langiella** (Hübn.) (*epilobiella* (Roem.) nec ([D. & S.]))
O. On a leaf of *Circaea lutetiana, Epilobium hirsutum* or *E. montanum*.
L. 6-7. In the leaves, making pale blotches; usually on plants in sheltered places, e.g. under trees or bushes.
P. 7-8. In a cocoon, amongst the leaves of the foodplant, or, sometimes, in the mined leaf.
I. 8-4. Fond of hibernating in thatch.

841(730). **M. terminella** (H. & W.)
O. On upperside of a leaf of *Circaea lutetiana*.
L. 8-9. When young, making spiral galleries; later, large pale blotches. Usually on plants growing in shady places in a wood.
P. 10-5. In a cocoon, sometimes in a mined leaf.
I. 6.

842(731). **M. locupletella** ([D. & S.]) (*schrankella* (Hübn.))
O. On *Epilobium alsinifolium, E. palustre, E. montanum* or *E. lanceolatum*.
L. 9-5;7. In the leaves, usually low on the plant, making blotches.
P. 5-6;7-8. In a white, spindle-shaped cocoon attached to grass or other vegetation.
I. 6;8. Possibly univoltine in the north.

843(732). **M. raschkiella** (Zell.)
O. 6;8. On the upperside of a leaf of *Epilobium angustifolium*, close to the midrib.
L. 6-7;9. In a long gallery leading to a blotch; readily changing leaves. The blotches when fresh are yellowish in colour; they bleach rapidly after the larvae have vacated them.
P. 7-8;?9-5. In a cocoon amongst detritus on the ground.
I. 5-6;8.

844(741). **M. miscella** ([D. & S.])
O. 6;9. On *Helianthemum* spp.
L. 7;10-4. In the leaves, making blotches.
P. 4-5;7-8. In a cocoon amongst detritus on the ground.
I. 5-6;8.

845(733). **M. conturbatella** (Hübn.)
O. On *Epilobium angustifolium*.
L. 5-6. In tightly spun leaves; in the crown of a plant before the stem has grown.
P. 6. In a greyish, spindle-shaped cocoon amongst detritus on the ground.
I. 6-7.

846(740). **M. ochraceella** (Curt.)
O. On *Epilobium hirsutum*.

L. 9-5. In the roots and stems, thence mining via the petiole into a lower leaf, where it makes a gallery on the underside; sometimes it mines a second leaf higher up on the plant.
P. 5-6. In a pale ochreous cocoon in the final mine; the leaf is contorted and the species is easiest to find in this stage.
I. 6-7.

847(735). **M. lacteella** (Steph.)
O. On *Epilobium hirsutum*.
L. 8. In a leaf, making a blotch.
P.
I. 6-7.

848(734). **M. propinquella** (Staint.)
O. 8. On *Epilobium hirsutum, E. montanum* or *E. palustre*.
L. 9-4. In a leaf, making a blotch.
P. 5. In a cocoon within the blotch.
I. 6-8.

849(736). **M. divisella** Herr.-Schäff. (*decorella* sensu auctt.)
O. 5-6. On *Epilobium montanum, E. palustre* or *E. lanceolatum*.
L. 6-7. In the stem, making a gall. Plants of *E. montanum* containing larvae are usually much branched above the gall, and the stems are redder.
P. 7-8. In a cocoon in the gall, with silk protruding from the exit-hole.
I. 8-5. Hibernates in thatch.

850(736A). **M. subdivisella** Bradley *Ent.Gaz.* **2**:173-182(1951)
O.
L.
P.
I. 9. Several specimens taken at Merton, Norfolk in 1894.

851(737). **M. nodicolella** Fuchs *Ent.Rec.* **63**:49(1951)
O. On *Epilobium angustifolium*.
L. 6-7. Feeds in the stem, making a gall or swelling, usually in the flowering part of the stem.
P. 7-8. In a thick silken cocoon spun amongst detritus on the ground.
I. 8-9.

852(738). **M. subbistrigella** (Haw.)
O. 5-6. On *Epilobium montanum*.
L. 7-8. In a seedpod. Pods containing larvae are usually thickened and shortened, sometimes distorted.
P. 8-9. Amongst detritus on the ground.
I. 9-5. Hibernates in thatch.

853(739). **M. epilobiella** ([D. & S.]) (*fulvescens* (Haw.))
O. 6. On *Epilobium hirsutum*.
L. 7. In a spun shoot of the foodplant.
P. 7-8. In the spun shoot, or a folded leaf nearby.
I. 8-5. Hibernates in thatch.

854(718). **Cosmopterix zieglerella** (Hübn.) (*eximia* (Haw.))
O. 7. On *Humulus*.
L. 8-5. In a silk-lined gallery along a leaf-vein, with clear branches where the larva has fed. When not feeding, the larva rests out of sight in the gallery. Overwinters in its cocoon.
P. 5-6. In old bine of hop.
I. 6-7.

855(719). **C. schmidiella** Frey
O. On *Vicia sepium*, often on plants in thick undergrowth.
L. 8-9. In a mine similar to that of the previous species.
P.
I. 6-7.

856(720). **C. orichalcea** Staint. (*druryella* sensu auctt.)
O. 7. On *Phalaris arundinacea, Phragmites, Milium* or *Hierochloe*.
L. 8-5. Making long, narrow, white galleries without frass in the leaves of the foodplant; can change leaves. Overwinters in its cocoon.
P. 5-6. In a rolled leaf on the ground.
I. 6.

857(721). **C. lienigella** L. & Z.
O. 7-9. On *Phragmites*.
L. 9-4. In a pale, pyriform blotch, with the frass packed at the narrow end; overwintering fully fed in the mine.
P. 5-6. In the blotch.
I. 6-7. In fens.

858(722). **Limnaecia phragmitella** Staint.
O. 8. On *Typha angustifolia* or *T. latifolia*.
L. 9-5. On the seeds, causing the down to hang out in large, conspicuous masses.
P. 6. In the seed-heads.
I. 7. Flies round the foodplant at dusk and later comes to light.

859(832). **Pancalia leuwenhoekella** (Linn.)
O. On *Viola hirta*.
L. 6-8. At first mining in a petiole, later in a silken gallery in the rootstock.
P.
I. 4-6. Flies in sunshine.

860(833). **P. latreillella** Curt.
O. On *Viola canina* or *V. hirta*.
L. 6-8. At first mining in a petiole, later in the rootstock.
P.
I. 4-5. Flies in sunshine.

861(752). **Euclemensia woodiella** (Curt.)
O.
L. In dead wood.
P.
I. 6. Taken in 1829. Probably now extinct in Britain.

862(729). **Glyphipteryx lathamella** (Fletch.) (*bimaculella* (Haw.) nec (Thunb.), nec (Schrank))
O. 7-8. On *Salix alba*.

L. 9-4. In the bark.
P. 5. In the bark.
I. 6-7.

863(728). **G. linneella** (Clerck)
O. 8-9. On *Tilia*.
L. 10-5. In the bark, extruding rust-coloured frass.
P. 5-6. In the bark.
I. 7-8. Rests on the trunks.

864(727). **Spuleria flavicaput** (Haw.) (*aurifrontella* (Hübn.))
O. 7-8. On *Crataegus*.
L. 8-10. In a twig.
P. In the twig. An elongate exit-hole on the underside, just beyond a fork, indicates the presence of a pupa.
I. 6. Flies over hawthorn in sunshine about 0900 hrs.

865(725). **Blastodacna hellerella** (Dup.) (*atra* sensu Ford)
O. 8-9. On *Crataegus*.
L. 9-10. In the berries. Tenanted berries have a small hole in the side.
P.
I. 6-7. Rests on the trunks and on fences; comes freely to light.

866(726). **B. atra** (Haw.) (*vinolentella* sensu Ford)
O. On *Malus*.
L. 4-5. In a shoot, usually a fruiting bud. Brown and shrivelled fruit buds and leaves at the end of a shoot indicate the feeding of a larva.
P. 5-6. In the shoot.
I. 7. Rests on the wood of the branches.

867(742). **Dystebenna stephensi** (Staint.)
O. Laid singly in a crevice of bark of *Quercus* spp.
L. In the living bark. Large and old trees are preferred.
P. In the bark.
I. 7-9. Rests on the trunks.

868(724 partim). **Sorhagenia rhamniella** (Zell.) *Ent.Gaz.* **14**:41-44(1963); *Ent. Rec.* **78**:9-11(1966)
O. On *Frangula* or **Rhamnus*.
L. 6. In spun shoots; feeding also on the flowers.
P. 6-7.
I. Mid 7-8.

869(724 partim). **S. lophyrella** (Dougl.) (*rhamniella* sensu Ford, partim) *Ent. Rec.* **78**:9-11(1966)
O. On *Rhamnus* or *Frangula*.
L. 4-5. In the buds and in loosely spun leaves.
P. 5-6.
I. Early 6.

870(724 partim). **S. janiszewskae** Riedl (*rhamniella* sensu Ford, partim) *Ent. Rec.* **78**:9-11(1966)
O. On *Frangula*.
L. 5-6. In a shoot, causing the terminal leaves to wither.
P. 6. In leaf litter.
I. End 6-7.

SCYTHRIDIDAE

871(889). **Scythris grandipennis** (Haw.)
O. On *Ulex*, preferring *U. minor*.
L. 2-4. In spun shoots.
P. 4-6. In the web.
I. 6.

872(890). **S. fuscoaenea** (Haw.)
O. On *Helianthemum* spp.
L. 5. On the leaves, from a silken web on the stems.
P. 5-6. In a silken cocoon on the ground.
I. 6-7.

873(892). **S. fallacella** (Schläg.) *Ent.* **68**:49-51(1935)
O. On *Helianthemum* spp.
L. 4-5. On the leaves, from a silken web on the stems.
P. 5. In a cocoon on the ground.
I. 5-6. Flies in evening sunshine. May be bivoltine, since the adult has been taken in August.

874(891). **S. fletcherella** Meyr.
O. On *Helianthemum* spp.
L. 5-6. On the leaves, from a silken web on the stems.
P. 6. In a cocoon on the ground.
I. 6-7.

875(893,894). **S. picaepennis** (Haw.) (*senescens* (Staint.); *vagabundella* (Herr.-Schäff.))
O. On *Lotus corniculatus* or *Thymus drucei*.
L. 5. On the leaves, from a silken web or tube which extends into the ground.
P. 6. In a cocoon in the ground.
I. 7. Males assemble to a virgin female. (894) *S. vagabundella*, which has a similar life history but is distinguished by its purple-fuscous forewings, occurs only on the west coasts of Scotland and Ireland, and is probably a form of this species.

876(895). **S. siccella** (Zell.)
O. On *Thymus drucei, Lotus corniculatus, Plantago* spp. and *Cerastium* spp.
L. 5. On the leaves from a long silken tube on or below the surface of sand, attached to half-buried stems of foodplant.
P. 6.
I. 7. Frequents flowers.

877(896). **S. empetrella** Karsholt & Nielsen (*variella* (Steph.))
O. On *Calluna* and *Erica* spp.
L. 4-5. On the leaves, from a long silken tube attached to twigs of the foodplant on or below the surface of the sand.
P. 5-6.
I. 6-7. Flies in sunshine.

878(897). **S. limbella** (Fabr.) (*quadriguttella* (Thunb.); *chenopodiella* (Hübn.))
O. On *Atriplex* and *Chenopodium* spp.
L. 4-5;7-8. In a web amongst the shoots and flowers.
P. 5-6;8-9.
I. 6;9. Rests on fences, etc.

879(898). **S. cicadella** (Zell.)
O. On *Scleranthus perennis*.
L. 5-6. In a tube of sand-coated silk amongst the roots and stems.
P. 6-7.
I. 7.

880(–). **S. potentillae** (Zell.) *Proc.Brit.ent.nat.Hist.Soc.*, **1978**: 11,14
O. *On *Rumex acetosa* or *R. acetosella*.
L. *4-5.
P. *6.
I. 6-7. Fond of resting in flowers of *Jasione montana* and *Convolvulus arvensis*.

COCHYLIDAE

881(268). **Hysterosia inopiana** (Haw.)
O. 8-9. On *Pulicaria dysenterica* or *Artemisia campestris*.
L. 9-4. In a silken tube in the roots.
P. 5-6.
I. 6-8. Rests well concealed by day. The male flies round the food-plants in the early evening, when the female may be found sitting on the leaves.

882(260). **H. schreibersiana** (Fröl.)
O. 6. On *Ulmus, Populus nigra* or *Prunus padus*.
L. 6-4. First on leaves spun together, then inside a shoot and leaf-stalk. Hibernates full-fed from September in a cocoon under the bark.
P. 4-5. In the cocoon.
I. 5-6. Sits high up in the trees except on emergence; has been taken lower down after heavy rain.

883(259). **H. sodaliana** (Haw.)
O. 6. On the green berries of *Rhamnus*.
L. 7-5. In the berries, spinning two or three together; a purple berry among green ones usually indicates a larva. Hibernates full-fed from late August in a cocoon in a crevice in the bark, near the ground.
P. 5-6. In the cocoon.
I. 6-7. Drops when disturbed by day. Flies for about half an hour towards dusk.

884(261). **Hysterophora maculosana** (Haw.)
O. 5-6. Singly on a seed-capsule of *Endymion non-scriptus*.
L. 7-4. In the seed-capsules, sealing the entrance-hole with silk. Hibernates full-fed from late August in the nearby stem of a dead umbellifer, etc.

P. 4-5. In the hibernaculum.
I. 5-6. The male flies in sunshine.

885(258). **Phtheochroa rugosana** (Hübn.)
O. 6. On the female flowers of *Bryonia*.
L. 6-4. First on spun flowers and immature berries; later inside a leaf spun to a stem, feeding on the latter. Hibernates full-fed from late September in a cocoon in the earth.
P. 4-5. In the cocoon.
I. 5-6. Inactive by day; flies towards dusk, later occasionally coming to light.

886(236). **Phalonidia manniana** (Fisch.v.Rösl.)
O. 7. On *Mentha aquatica* or *Lycopus*.
L. 8-4. In the stem, overwintering full-fed therein from October.
P. 5-6. In the stem, in a cocoon of silk and frass; emergence-hole not visible.
I. 6-7. Flies freely in the evening and at night, when it comes to light.

887(238). **P. minimana** (Caradja) (*walsinghamana* (P. & M.))
O. 6-7;8-9. On *Pedicularis palustris*.
L. 7;9-4. In the seed capsules.
P. 5;7. In a strong silken cocoon amongst surface debris or in the earth.
I. 5-6;7-8. Difficult to disturb; makes short flights round the food-plant on warm evenings.

888(239). **P. permixtana** ([D. & S.]) (*mussehliana* (Treits.))
O. *On *Butomus, Pedicularis, Alisma plantago-aquatica, Euphrasia* or *Rhinanthus*.
L. *6-7;9-4. In the flower-heads, seeds and stem.
P.
I. 5-6;7-8. Frequents rough, damp ground with sparse vegetation. Flies readily if disturbed by day; active in the evening and later comes to light. Early stages unrecorded in Britain.

889(234). **P. vectisana** (H. & W.)
O. 6;9. On *Triglochin maritima* or *T. palustris*.
L. 6;9-4. In June on the flowers; second generation in the shoots just above the crown, mining downwards into the roots, where they overwinter.
P. 4-5;6-7. In a silken cocoon in a dead flower-stem, or amongst ground debris.
I. 5-6;7-9. Frequents saltings. Seldom flies if disturbed by day, but does so freely, often in large numbers, in the evening sunshine.

890(237). **P. alismana** (Rag.)
O. 6-8. On the upper part of a flower-stem of *Alisma plantago-aquatica*.
L. 9-4. In a flower-stem; hibernates full-fed from October in a grey, silken cocoon in the lower part of the stem.
P. 5. In the cocoon. Stems gathered in the winter will usually produce imagines.
I. 6-8. Prolonged emergence period. Flies freely at sunset.

891(235). **P. luridana** (Gregs.)
O.
L. Once bred from a flower-head of *Matricaria recutita*, gathered in September of the previous year.
P.
I. 5-8, but possibly bivoltine. Rests by day amongst herbage and on flower-heads of *M. recutita*; flies at sunset and later comes to light.

892(233). **P. affinitana** (Dougl.)
O. 6-7. On a flower of *Aster tripolium*.
L. 7-3. At first on the flowers, then boring down the stem into the rootstock, where it overwinters full-fed from late October.
P. 3-5. In the rootstock or nearby tidal debris.
I. 6-8. Flies occasionally on sunny afternoons, but the main flight-time is late evening.

893(254). **P. gilvicomana** (Zell.)
O. 6-7. On a flower-bud or stem of *Mycelis* or *Lapsana*.
L. 7-5. On the seeds, generally concealed within the pappus. Hibernates full-fed from August in a slight silken cocoon amongst leaf-litter or in the soil.
P. 5. In the cocoon.
I. 6-7. Flies in the evening at about head-height.

894(253). **P. curvistrigana** (Staint.)
O. 7-8. On a late green bud or still-closed seed-head of *Solidago virgaurea*.
L. 8-4. On the unripe seeds, sometimes spinning the flowers together; an unopened seed-head may indicate the presence of a larva. Hibernates full-fed from September in a silken cocoon on the ground.
P. 4-6. In the cocoon.
I. 7-8. Seldom seen by day; flies in the evening and at sunset.

895(264). **Stenodes alternana** (Steph.)
O. 8. On the crown of *Centaurea scabiosa*.
L. 9-7. Overwinters small in the crown. In spring feeds in the developing flower-buds; frass sticking to the base of the bud indicates that a large larva is at work.
P. 7. In the bud.
I. 7-8. Flies over the foodplant at dusk.

896(263). **S. straminea** (Haw.)
O. 7;9. On *Centaurea nigra*.
L. 7-8;9-5. In summer, in the upper stem under the flower-head, eating the young seeds. Autumn larvae hibernate in the crown when still small; in spring they feed in the stems of shoots, turning the tips brown.
P. 4-5;7-8. In a cocoon in earth or debris in spring; in the larval habitation in summer.
I. 5-7;8-9. Rests on the foodplant by day; flies from dusk onwards, coming to light.

897(267). **Agapeta hamana** (Linn.)
O. 6-9. On *Carduus* spp.; possibly also on *Cirsium* spp. Oviposition is spread over a long period.
L. 8-4. In the roots. Hibernates full-fed from October in a silken cocoon spun in the larval habitation or the soil.
P. 5. In the cocoon.
I. 5-9. Apparently univoltine with a very long emergence period. Readily disturbed by day. Flies in the evening and later comes to light.

898(266). **A. zoegana** (Linn.)
O. 6-8. On *Centaurea nigra* or *Scabiosa*. Oviposition is spread over a long period.
L. 9-4. In the roots, overwintering there full-fed from October in a silken cocoon.
P. 4-5. In the cocoon.
I. 5-8. Has a prolonged emergence period. Flies from sunset onwards.

899(232). **Aethes tesserana** ([D. & S.])
O. 6-8. On *Picris hieracioides, P. echioides, Hieracium, Crepis* or *Inula conyza*.
L. 9-4. In the rootstock and roots, giving no external evidence of feeding.
P. 5-6.
I. 5-8. Has a prolonged emergence period. Flies in sunshine.

900(225). **A. rutilana** (Hübn.)
O. 7-8. Laid singly in a leaf-axil of *Juniperus*.
L. 9-6. First mines the leaves, ejecting frass; overwinters in a frass-covered tube spun against a twig; in spring, eats the concave surface of the leaves from a web.
P. 7. In a slight cocoon in the larval habitation.
I. 7-8. Flies close to the ground in sunshine during mid-day hours. Best taken by beating juniper over a tray in the early morning.

901(255). **A. hartmanniana** (Clerck) (*subbaumaniana* (Staint.))
O. Probably on *Knautia* or *Scabiosa*.
L. Probably on the rootstocks.
P. 4-5. Probably in the larval habitation.
I. 6-8. A downland species. It is easily disturbed by day and flies freely at dusk.

902(256). **A. piercei** Obraztsov (*baumanniana* sensu auctt.)
O. 7-8. On *Succisa*.
L. 8-4. In the rootstock.
P. 4-5. In the rootstock.
I. 5-7. Frequents rough damp habitats. Is easily disturbed by day and flies freely at dusk.

903(228). **A. margarotana** (Dup.) (*maritimana* (Guen.))
O. 7. In a leaf-axil or just below a flower of *Eryngium maritimum*; hatches in 12 to 14 days.
L. 7-4. At first in the upper stem; later in the roots, hollowing them out and causing black discoloration.

P. 4-5. In a pupal chamber in the rootstock.
I. 6-7. Seldom flies, but may be found resting on or near the food-plant. May be bred by potting up about 12 cm of root and 7 cm of stem, gathered in the winter.

904(227). **A. williana** (Brahm) (*zephyrana* (Treits.))
O. 6-8. On *Daucus*.
L. 8-4. In the lower stem, exuding frass in the autumn; later mining downwards into the rootstock.
P. 4-6. In the rootstock.
I. 5-8. Flies in the evening and at sunset, earlier in warm, dry weather.

905(231). **A. cnicana** (Westw.)
O. 6-8. On a flower of *Cirsium* spp.
L. 9-4. First eating the seeds; later boring into the stem and eating the pith. Hibernates full-fed in the stem.
P. 5-6. In the stem.
I. 5-7. Easily disturbed. Flies in the evening and later comes to light.

906(230). **A. rubigana** (Treits.) (*badiana* (sensu Hübn.))
O. 7. On a young burr of *Arctium lappa*; possibly also on *A. minor*.
L. 9-5. On the seeds. Hibernates full-fed from October in a cocoon on the ground or, occasionally, in the seed-head. Seed-heads collected in September and overwintered out-doors often yield moths.
P. 5-6. In the cocoon.
I. 6-8. Flies in the late evening and later comes to light.

907(229). **A. smeathmanniana** (Fabr.)
O. 6-8. On *Achillea millefolium*, *Centaurea* spp. or *Anthemis arvensis*.
L. 9-5. In spun flowers or in the seed-head, overwintering in the feeding-place. Spinnings on *Achillea* are visible throughout the winter.
P. 5. In a silken cocoon amongst detritus.
I. 5-8. Easily disturbed on warm days. Flies freely in the early evening.

908(226). **A. margaritana** (Haw.) (*dipoltella* (Hübn.))
O. 7-8. On a flower-head of *Achillea millefolium*, *Matricaria recutita* or *Leucanthemum vulgare*.
L. 9-4. On the flowers and seeds, hibernating full-fed in a silken spinning in the feeding-place.
P. 5-6. In a cocoon spun amongst detritus on the ground.
I. 7-8. Easily disturbed in the daytime; flies in the evening and at dusk.

909(224). **A. dilucidana** (Steph.)
O. 7-8. Laid singly on a flower of *Pastinaca*.
L. 8-4. On the seeds. In September, it bores into the stem and hibernates there full-fed; whitish powder at a leaf-axil shows where the larva entered.

P. 5-6. In the stem.
I. 7-8. Easily disturbed from the foodplant in the daytime. Flies towards sunset. Mainly coastal.

910(223). **A. francillana** (Fabr.)
O. 7-9. On *Daucus* or, less often, *Peucedanum officinale*.
L. 9-5. In spun flowers and on the seeds; in the autumn, boring into the lower part of the stem, where it overwinters. In spring, it returns to the upper part of the stem. Several larvae may be found in one stem, their presence being indicated by an exudation of white frass.
P. 6-7. In the stem, just below the entrance-hole.
I. 6-9. Easily disturbed from the foodplant in daytime; flies from sunset onwards.

911(222). **A. beatricella** (Wals.)
O. 7. On *Conium* or, occasionally, *Smyrnium*.
L. 8-4. On the seeds, then entering the stem to hibernate, often gregariously.
P. 5. In the stem.
I. 6-7. Flies from late evening to sunset.

912(265). **Commophila aeneana** (Hübn.)
O. 6-8. On *Senecio jacobaea*.
L. 9-4. In the rootstock. Hibernates in the lower part of the stem, which breaks off above the burrow.
P. 5. In the hibernaculum.
I. 5-7. Flies close to the ground towards sunset, or rests on the foodplant.

913(–). **Eugnosta lathoniana** (Hübn.)
Two or three specimens were allegedly taken in the early part of the nineteenth century.

914(262). **Eupoecilia angustana** (Hübn.)
O. 7-9. On a flower of *Plantago, Achillea, Origanum, Solidago virgaurea* or *Thymus*; f. *fasciella* Don. on *Calluna*.
L. 7-4. On spun flowers and seeds, hibernating full-fed from October in the spinning.
P. 4-5. In the spinning.
I. 6-9. Has a prolonged emergence period; flies at dusk. f. *fasciella*, which occurs on heathland, flies over the heather on warm afternoons.

915(257). **E. ambiguella** (Hübn.)
O. 6. On a flower of *Frangula*; less often on *Hedera, Lonicera* or *Swida*.
L. 7-4. In the berries, spinning two or three together. When full-fed in September it constructs a leaf-case amongst detritus on the ground or attached to a stem, overwintering therein.
P. 4-5. In the leaf-case; a few larvae pupate in the autumn.
I. 5-6. May be beaten out, more easily late in the day; flies freely towards dusk.

916(242). **Cochylidia implicitana** (Wocke)
O. 6-8. On *Matricaria recutita, Anthemis cotula, Solidago virgaurea* or *Tripleurospermum maritimum* subsp. *inodorum*.
L. 6-4. On the flowers and seeds or in a stem or shoot, eating the pith; overwinters full-fed.
P. 4-6. In the larval feeding place or amongst detritus on the ground.
I 5-8. Apparently univoltine with a long emergence period, though Ford (1949) gave three generations. More easily disturbed in late afternoon; flies towards sunset and later comes to light.

917(247). **C. heydeniana** (Herr.-Schäff.) (*sabulicola* (Wals.))
O. 6;8. On *Erigeron acer* or *Conyza canadensis*.
L. 6-7;8-4. First generation in the seed-heads, those which contain larvae being compressed at the tip with some seed-down protruding. Second generation in the central shoot, eating the seeds; hibernates full-fed from September in the feeding place.
P. 4-5;7. In a slight cocoon in the feeding place.
I. 6;7-8. Easily disturbed by day; flies actively in the evening.

918(241). **C. subroseana** (Haw.)
O. 7. Laid singly on a flower of *Solidago virgaurea*.
L. 8-4. On the flowers and seeds, retarding their development. Hibernates full-fed from late September in a cocoon amongst detritus on the ground.
P. 5-6. In the cocoon.
I. 6-8. Flies rather slowly in the evening.

919(240). **C. rupicola** (Curt.)
O. 7. On *Eupatorium*.
L. 8-4. On the flowers and seeds. Hibernates full-fed from October in a broken stump of the foodplant or a piece of dead stem lying on the ground; these may be collected during the winter.
P. 5-6. In a cocoon in the dead stump or prostrate stem.
I. 6-7. Flies in warm afternoon sunshine and again at dusk.

920(245). **Falseuncaria ruficiliana** (Haw.) (*ciliella* (Hübn.))
O. 6-7. On *Primula veris* or *Pedicularis sylvatica*.
L. 7-4. In the seed-capsules. Hibernates full-fed from September in a cocoon spun amongst ground detritus or in rotten wood.
P. 4-5. In the cocoon.
I. 5-6. Can be disturbed by day.

921(244). **F. degreyana** (McLach.)
O. 6;8. On *Linaria vulgaris* or *Plantago lanceolata*.
L. 6;8-9. The first generation on the flowers; the second on the seeds.
P. 5;7. In the larval habitation on *Plantago*; or amongst detritus or in the soil.
I. 5-6;7-8. Flies towards sunset.

922(243). **Cochylis roseana** (Haw.)
- O. 7-8. Laid singly on a seed-head of *Dipsacus fullonum*.
- L. 8-5. On the seeds, tunnelling from one to another, but never entering the central cavity (cf. no. 1059).
- P. 5-7. In the seed-head.
- I. 5-8. Flies in the afternoon sunshine and towards dusk.

923(246). **C. flaviciliana** (Westw.)
- O. 7-8. In a flower-head of *Knautia*.
- L. 7-4. In the seed-head. Hibernates full-fed from October in a cocoon spun amongst detritus on the ground.
- P. 5-7. In the cocoon.
- I. 6-8. Seldom seen by day; flies round the foodplant at dusk.

924(251). **C. dubitana** (Hübn.)
- O. 6-7;8-9. On *Senecio jacobaea*, *Crepis* spp., *Solidago virgaurea*, *Hieracium* or other Compositae.
- L. 7;9-4. On the flowers and in the seed-head; the second generation hibernates full-fed from October in a cocoon.
- P. 4;7. In the cocoon spun in both generations amongst detritus.
- I. 6;8. Can be disturbed in late afternoon and flies freely in the evening.

925(252). **C. hybridella** (Hübn.)
- O. 8. On *Picris hieracioides*, *P. echioides* or *Crepis* spp.
- L. 8-4. On the flowers and unripe seeds. Hibernates full-fed from October in a cocoon spun amongst detritus on the ground.
- P. 5. In the cocoon.
- I. 7-8. Flies freely at sunset.

926(249). **C. atricapitana** (Steph.)
- O. 6-7;8-9. On *Senecio jacobaea*.
- L. 7;9-4. First generation in the flowers, or flower-stalks; second generation in the lower stem and rootstock, where it over-winters full-fed from October.
- P. 4;7. In a white silken cocoon in the larval feeding place.
- I. 5-6;7-8. The male flies at sunset, the female later; both come to light.

927(250). **C. pallidana** Zell.
- O. 6-7. On *Jasione*.
- L. 7-4. On the flowers or in the seed-head; hibernates full-fed from September in a cocoon spun amongst detritus on the ground.
- P. 4-5. In the cocoon.
- I. 6. Rests concealed by day; flies at dusk.

928(248). **C. nana** (Haw.)
- O. 6. On the catkins of *Betula* spp.
- L. 7-4. In a catkin; hibernates full-fed from October in a cocoon spun in a bark crevice.
- P. 5-6. In the cocoon.
- I. 6. Rests in the daytime on trunks and fences, and is easily disturbed; flies round birches at dusk.

TORTRICIDAE

TORTRICINAE

929(284). **Pandemis corylana** (Fabr.)
O. On *Corylus, Fraxinus, Prunus* spp., *Rubus* spp., *Quercus* spp.
L. 5-7. In leaves spun together, or folded longitudinally.
P. 6-7. In the larval habitation.
I. 7-9. Easily disturbed from its daytime hideout amongst foliage; flies at dusk, occasionally coming to sugar.

930(287). **P. cerasana** (Hübn.) (*ribeana* (Hübn.))
O. Laid in small batches on leaves or branches. Some hatch in late summer, some in spring. On *Acer, Alnus glutinosa, Corylus, Quercus* spp., *Salix* spp., *Vaccinium* spp., *Tilia, Betula* spp., *Ulmus, Lysimachia, Sorbus*, and all fruit trees.
L. 9-5. In a folded or rolled leaf.
P. 5-6. In a whitish cocoon in a folded leaf, or the larval habitation.
I. 6-8. Habits similar to those of *P. corylana*.

931(285). **P. cinnamomeana** (Treits.)
O. On *Acer, Abies, Betula* spp., *Larix, Picea, Prunus* spp., *Sorbus, Taxus* or *Vaccinium* spp.
L. 5-6. In spun leaves.
P. 6-7. In the larval habitation.
I. Late 6-7. Readily disturbed by day from foliage and bracken when it darts to the ground in a zigzag flight; comes to light.

932(286). **P. heparana** ([D. & S.])
O. In batches of 30-50 on upper leaf-surfaces.
L. 5-6. In a rolled leaf. Polyphagous on trees and shrubs.
P. 6-7. In the larval habitation or spun leaves, usually near a shoot tip.
I. 6-8. Easily put up by day; flies at dusk and later comes to light.

933(297). **P. dumetana** (Treits.)
O. 8. In batches of variable number on the underside of leaves.
L. 9-6. In a rolled leaf; occasionally in a flower head. Polyphagous on herbaceous and other plants.
P. 6-7. In the larval habitation.
I. 7-8. Flies from sunset onwards.

934(302). **Argyrotaenia pulchellana** (Haw.) (*politana* (Haw.))
O. 5;8. In batches of 40-50, usually on the upper surface of leaves of many plants, but especially on *Erica, Calluna, Myrica* or *Vaccinium* spp.
L. 6;9. Gregarious when young, skeletonizing the underside of a leaf along the midrib; then singly in spun leaves.
P. 6;10-4. In a cocoon in a spun leaf or, for the second generation, in detritus on the ground.
I. 4-5 ;6-7, but univoltine in the north, flying late 4-6. Rests on its foodplants by day but flies freely in the late afternoon.

935(–). **Homona menciana** (Walk.)
A tropical species which has been bred from imported *Camellia* plants.

936(274). **Archips oporana** (Linn.) (*piceana* (Linn.))
O. 8. On *Pinus sylvestris, Abies alba, Picea glauca* or *Juniperus*.
L. 9-6. Overwinters in a silken tube spun among the needles, then eats the new shoots, and bores into the stem. On *Juniperus*, in a thick white web.
P. 6-7. In the larval habitation.
I. 6-7. May be disturbed by tapping branches of large pines early in the day. Flies from mid-afternoon to early evening and at dusk around conifers.

937(275). **A. podana** (Scop.)
O. 7. In batches of 50-100 laid on upper leaf surfaces under a wax-like secretion.
L. 7-5. Polyphagous on trees and shrubs; sometimes an orchard pest. Larva in spun leaves and flowers.
P. 6-7. In the larval habitation.
I. 6-7. Sits on the leaves by day, but flies freely in the late afternoon and at dusk.

938(276). **A. betulana** (Hübn.) (*decretana* (Treits.))
O. On *Myrica*.
L. 5-7. In spun leaves of a terminal shoot.
P. 7. In a slight web in the larval habitation, or leaf-litter on the ground.
I. 8. Not easily disturbed from its hiding place among foliage by day; flies swiftly over the foodplant from dusk. Not recorded since 1900.

939(277). **A. crataegana** (Hübn.)
O. 8-4. In batches on the bark, under a white gum looking like a bird-dropping, of *Quercus* spp., *Ulmus, Fraxinus, Tilia, Salix* spp., orchard or other trees.
L. 4-5. In a leaf-edge rolled into a tight tube.
P. 6-7. In the larval habitation.
I. 6-8. Flies fast when disturbed by day, and from sunset freely and erratically about the trees.

940(278). **A. xylosteana** (Linn.)
O. In small batches under a brownish secretion on the bark of deciduous trees, *Abies, Rubus, Lonicera* or *Hypericum*.
L. 4-6. In a rolled leaf.
P. 6. In the larval habitation.
I. 7. Similar behaviour to *A. crataegana*.

941(279). **A. rosana** (Linn.)
O. 8-4. In greenish masses, usually on the bark of fruit and other deciduous trees and shrubs; occasionally on conifers or *Myrica*.
L. 4-6. First in a bud, then in rolled or spun leaves; also on flowers and fruits.
P. 6-7. In the larval habitation, in a cocoon.
I. 7-9. Readily disturbed by day. Flies from dusk.

942(298). **Choristoneura diversana** (Hübn.)
O. 8-9. In large batches on various trees, shrubs, and herbaceous plants.

L. 8-5. In spun leaves.
P. 6. In the larval habitation, or freshly spun leaves.
I. 7. Drops when disturbed by day; flies at dusk and at dawn over the tops of young trees.

943(280). **C. hebenstreitella** (Müll.) (*sorbiana* (Hübn.))
O. 8. On *Quercus* spp., *Betula* spp., *Corylus, Salix* spp., *Myrica, Vaccinium* spp., *Hedera*, etc.
L. 9-5. In spun or rolled leaves.
P. 6. In the larval habitation.
I. 6-7. Not easily disturbed by day; flies wildly from dusk onwards.

944(281). **C. lafauryana** (Rag.)
O. 7-8. In batches on the upper surface of leaves of *Myrica*.
L. 8-7. In a vertical tube spun from the upper terminal leaves.
P. 7. In the larval habitation, or spun up in detritus on the ground.
I. 7-8. The male flies when disturbed; the female drops to the ground. Flight period from sunset onwards; occasionally comes to light.

945(283). **Cacoecimorpha pronubana** (Hübn.)
O. In small greenish batches on the upperside of leaves of a variety of plants, particularly on *Euonymus japonicus*; sometimes a garden or greenhouse pest.
L. 5-8;9-4. In spun leaves or seeds. A few first-brood larvae overwinter.
P. 4-6;8-10. In spun or folded leaves.
I. 5-7;8-11. The broods overlap. Flies during the day in sunshine with a characteristic erratic flight.

946(299). **Syndemis musculana** (Hübn.)
O. 7. Usually laid on *Betula* spp., *Rubus* or *Quercus* spp., less often on grasses or herbaceous plants.
L. 7-4. In a compact tube of spun leaves, or folded leaves, hibernating therein.
P. 4-5. In the larval habitation or ground-litter.
I. 5-7. Easily disturbed by day; flies from late afternoon till dark.

947(283A). **Ptycholomoides aeriferanus** (Herr.-Schäff.)
O. On *Larix*.
L. 5-6. In spun needles.
P. 6. In the larval habitation.
I. 6-8. Seldom seen by day; mostly seen at light.

948(291). **Aphelia viburnana** ([D. & S.])
O. 7. On *Vaccinium, Erica, Myrica*, etc.
L. 8-6. In spun leaves.
P. 6-7. In a white silken cocoon where the larva has fed.
I. 7. Easily put up by day and flies actively on hot, sunny days. Two races occur, one on heathland, the other on saltings.

949(290). **A. paleana** (Hübn.)
O. 8-9. Laid in small batches two by two along the upperside of a leaf of various grasses and herbaceous plants; occasionally on *Quercus* spp. or *Fagus*.
L. 9-6. In spun leaves.

P. 6. In spun leaves.
I. 7-8. Easily disturbed by day; flies slowly over rough herbage in the evening.

950(–). **A. unitana** (Hübn.)
O. On *Heracleum, Rubus, Angelica, Allium*, and probably other herbaceous plants.
L. 5-6. In spun leaves.
P. 6. In the larval habitation.
I. 6-7. Frequents high moorland and limestone dales.

951(292). **Clepsis senecionana** (Hübn.) (*rusticana* sensu Treits. et auctt.)
O. 6-7. On *Myrica* or *Vaccinium myrtillus*; also *Picea* or *Pinus*; on *Larix* in Scotland.
L. 7-4. In a rolled leaf, or spun terminal shoot.
P. 4. In a white silken cocoon in a spun shoot.
I. 5-6. Not easily aroused by day, but flies freely in warm sunshine from afternoon onwards, disappearing at the onset of clouds.

952(295). **C. rurinana** (Linn.) (*semialbana* (Guen.))
O. On *Rosa, Lonicera, Acer, Fagus* or *Quercus* spp.
L. 4-6. In rolled leaves.
P. 6. In the larval habitation.
I. 6-7. Retiring in its habits; drops when disturbed. Flies at dusk.

953(294). **C. spectrana** (Treits.) (*costana* sensu Fabr. et auctt.)
O. Laid in small batches on *Epilobium* spp., *Filipendula, Potentilla* and most marshland plants; on *Limonium vulgare, Artemisia maritima* and other maritime plants.
L. 5-6. In spun leaves.
P. 5-6. In a white silken cocoon in the spinning.
I. 5-7;8-9. There is a small second brood from the earliest adults. Not easily disturbed by day; flies gently towards dusk, and later comes to light.

954(296). **C. consimilana** (Hübn.) (*unifasciana* Dup.)
O. 7-9. Laid in small batches along the upper midrib of a leaf of *Ligustrum*, or, less frequently, of *Syringa, Hedera, Lonicera, Polygonum* spp., *Malus, Carpinus* or *Crataegus*.
L. 8-6. In dense, untidy spinnings, preferring dead or withered leaves. Hibernates in the third instar.
P. 6. In the feeding place.
I. 6-9. Readily disturbed by day; flies freely in the afternoon and evening and later comes to light.

955(–). **C. trileucana** (Doubl.)
A North American species, two of which were allegedly taken in Cornwall early in the nineteenth century.

956(–). **C. melaleucanus** (Walk.)
A North American species, two of which were allegedly taken in the West of England early in the nineteenth century.

957(–). **Epichoristodes acerbella** (Walk.)
An African species occasionally imported accidentally as a larva on carnations (*Dianthus*).

958(300). **Epiphyas postvittana** (Walk.)
O. 6;8-9. Laid in small batches on the upperside of a leaf of *Euonymus japonicus* or some other shrub or plant, particularly those in gardens.
L. 6-7;9-4. In loosely spun shoots.
P. 5-6;8. In the larval habitation.
I. 5-10, the broods overlapping. Easily disturbed by day; flies freely after dusk and comes to light.

959(273A). **Adoxophyes orana** (Fisch.v.Rösl.)
O. 6;8-9. Laid in batches on the leaves or, occasionally, the fruits of *Malus, Pyrus, Prunus* and a wide range of other trees and plants.
L. 6-7;9-5. Scarring the fruit when small; later in spun leaves.
P. 5-6;8-9. In the larval habitation or freshly spun leaves.
I. 6;8-10. The broods may overlap. Readily disturbed by day; flies towards dusk and later comes to light. An orchard pest.

960(282). **Ptycholoma lecheana** (Linn.)
O. 7-8.
L. 8-5. In rolled leaves of various trees and shrubs, including fruit trees.
P. 6. In a dense white cocoon in a rolled leaf, or the larval habitation.
I. 6-7. Flies wildly over trees and bushes in the afternoon and evening.

961(303A). **Lozotaeniodes formosanus** (Gey.)
O. 7-8. On a needle of *Pinus sylvestris*, in a row; hatching in 14-21 days.
L. 9-6. When young, on the brown scaling of the buds; after hibernation, in an open-ended silken tube among the needles.
P. 6-7.
I. 6-8. Seldom seen by day. Flies at dusk and comes readily to light, often far from its foodplant.

962(293). **Lozotaenia forsterana** (Fabr.)
O. 8-9. On *Hedera, Ligustrum, Lonicera, Stachys, Larix, Abies, Picea* or *Pinus*.
L. 9-5. Between two leaves spun together.
P. 6. In spun leaves, or the larval habitation.
I. 6-7. Frequents gardens, hedgerows, copses and wood-borders.

963(–). **L. subocellana** (Steph.)
Known only from two specimens (now lost), allegedly taken in Devon or Cornwall early in the nineteenth century.

964(271A). **Paramesia gnomana** (Clerck) *Ent.Rec.* **89**:274(1977)
O.
L. *5-6 Polyphagous, feeding on the leaves of its foodplant.
P.
I. 6-8. Frequents damp places, hiding by day and is not easily disturbed; flies at night and comes to light.

965(303). **Periclepsis cinctana** (*Paraclepsis cinctana*) ([D. & S.])
O. *On *Anthyllis, Lotus, Genista, Sarothamnus* or *Artemisia campestris*.
L. *9-6. In a tubular silken gallery.
P.
I. 7. Frequents dry grassland on chalk downs and sea-cliffs. Flies close to the ground for short distances in sunshine.

966(271). **Epagoge grotiana** (Fabr.)
O. 7-8. Laid in batches of two or three on *Quercus* spp., *Crataegus* or *Rubus* spp.
L. 8-5. On the leaves. Overwinters in a narrow fold of a leaf.
P. 6. In a spinning in or near the larval habitation.
I. 6-7. Not easily disturbed by day; active at dusk and dawn and sometimes comes to light.

967(270). **Capua vulgana** (Fröl.) (*favillaceana* (Hübn., 1814-17), nec (Hübn., 1796-99))
O. On *Carpinus, Alnus glutinosa, Sorbus* or *Vaccinium* spp.
L. 7-9 or 7-4. Feeds on the leaves.
P. 9-5 or 4-5. Pupation may take place before or after hibernation.
I. 5-6. The male flies by day, and both sexes towards dusk; rests low down on trunks.

968(273). **Philedone gerningana** ([D. & S.])
O. On *Potentilla, Helianthemum, Vaccinium* spp. or *Armeria maritima*; also *Picea sitchensis* in Scotland.
L. 5-6. In spun leaves and flowerheads.
P. 6. In the larval habitation.
I. 7-9. The male flies at any time when the sun shines, the female only in the evening.

969(272). **Philedonides lunana** (Thunb.) (*prodromana* (Hübn.))
O. 3-5. In small batches under a gum on the leaves of *Potentilla, Mentha, Myrica, Vaccinium* spp., *Erica, Calluna, Smyrnium* etc.; on some conifers in Scotland and Wales.
L. 5-6. In spun or folded leaves or flowers.
P. 6-3. In spun leaves of the foodplant or detritus on the ground.
I. 3-5. The male flies freely in calm, sunny weather especially at sunrise, midday and in the evening; the female sits on vegetation.

970(269). **Ditula angustiorana** (Haw.)
O. 6-7. On leaves of many trees, shrubs and sometimes herbaceous plants.
L. 8-5. Among spun leaves and fruit buds.
P. 6. Usually on the foodplant in spun or folded leaves, or in detritus on the ground.
I. 6-7. The male is active in sunshine and the female is easily disturbed; both sexes also fly at dusk. An orchard pest.

971(325). **Pseudargyrotoza conwagana** (Fabr.) (*conwayana* (Gmelin))
O. 7-8. On *Ligustrum* or *Fraxinus*.
L. 8-10. In the berries and seeds.
P. 10-4. In a flattened white cocoon on the ground.
I. 5-7. Flies actively round the trees in warm sunshine.

972(327). **Sparganothis pilleriana** ([D. & S.])
O. 7. In batches of 40-60 on a wide range of plants, especially those characteristic of salt-marshes and inland bogs.
L. 9-6. In spun or folded leaves, usually folded from the upperside.
P. 6. In a spun or rolled leaf.
I. 7. Seldom seen by day; the male flies actively just before dark.

973(322). **Olindia schumacherana** (Fabr.) (*ulmana* (Hübn.))
O. Principally on *Ranunculus ficaria*; also on *Aquilegia, Mercurialis, Vaccinium* spp., etc.
L. 4-6. In a downward-folded leaf, eating the lower cuticle but leaving the upper surface intact.
P. 6. In the larval habitation, or a cocoon on the ground.
I. 6-7. Hides by day in thick cover; easily disturbed but hides again quickly. The male flies in the afternoon, the female at dusk.

974(321). **Isotrias rectifasciana** (Haw.)
O. 6-7. Laid in small batches under a secretion.
L. 7-?. In captivity, larvae fed on *Crataegus* and *Ulmus* from a silk tube on the underside of a leaf. They went into hibernation in mid-September but escaped during the winter; mortality had been high (J.M. Chalmers-Hunt). The feral foodplant is unknown.
P.
I. 5-6. Easily beaten from hedges, but does not fly far. Flies freely at sunset.

975(301). **Eulia ministrana** (Linn.)
O. 5-6. In small batches on various trees and shrubs; also on *Vaccinium myrtillus*.
L. 7-4. Between two leaves spun as an open tube. Hibernates full fed.
P. 4-5. In the tube with the openings closed.
I. 5-6. Seen little by day; flies fairly high about trees at sunset.

976(308). **Cnephasia longana** (Haw.)
O. 7. Laid singly or in small batches on a wide range of herbaceous plants.
L. 8-6. The newly hatched larva hibernates before feeding; in spring it feeds in spun flowers and terminal shoots.
P. 6. In a flimsy cocoon on the ground, in dead leaves or a seed capsule.
I. 7. Readily disturbed by day. Flies from dusk onwards.

977(–). **C. gueneana** (Dup.) *Ent.Gaz.* **12**:18(1961)
O.
L. Polyphagous on herbaceous plants; has been reared from imported *Anemone*.
P.
I. A Mediterranean species which has been imported with garden plants.

978(313). **C. communana** (Herr.-Schäff.)
O.
L. Early stages undescribed; said to be polyphagous on herbaceous plants.
P.
I. 5-6. Sits on tree-trunks, fences and in vegetation in the daytime; easily disturbed. Flies early in the morning and at night.

979(318). **C. conspersana** Dougl.
O. Laid in small batches on a wide range of foodplants, but principally on Rosaceae or Compositae.
L. 6. In the flowers, spinning together the petals, or in spun leaves of a terminal shoot, then eating out the heart.
P. 6-7. Usually in detritus on the ground.
I. 7. In open situations e.g. cliffs and downs. The male often sits on posts etc., and may easily be put up in sunny weather; both sexes fly at night.

980(315,314). **C. stephensiana** (Doubl.) (*chrysantheana* sensu auctt.)
O. 8-9. Laid singly or in small batches on a wide range of herbaceous plants.
L. 9-5. On hatching, hibernates without feeding. In the spring, first mining a leaf, then in spun leaves, the type of spinning varying according to the plant.
P. 6-7. In the larval habitation, or spun up nearby on the plant.
I. 7-8. Easily disturbed by day; flies at early dusk and later comes to light. (314) f. *octomaculana* Curt., formerly regarded as a distinct species, has a similar life history.

981(312). **C. interjectana** (Haw.) (*virgaureana* (Treits.))
O. Laid singly or in batches of two or three on a wide range of herbaceous plants.
L. 9-6. In the first instar mining a leaf; after hibernation in spun leaves and flowers.
P. 6. In the folded edge of a leaf, or in detritus on the ground.
I. 6-8. Commonly rests exposed by day. Easily disturbed but flies quickly to cover. Flies from dusk onwards.

982(316). **C. pasiuana** (Hübn.) (*pascuana* (Hübn.))
O. On Compositae or *Ranunculus*.
L. 5-6. On the flowers, spinning the florets together.
P. 6. Under the folded edge of a leaf, or in detritus on the ground.
I. 6-7. Easily beaten out by day; flies in the evening.

983(317). **C. genitalana** P. & M.
O. On *Senecio, Leucanthemum, Ranunculus, Centaurea, Anthyllis* or *Mentha* spp.
L. 6. In spun flowers and leaves.
P. 6-7. In a flimsy cocoon amongst detritus on the ground.
I. 7-8. On chalk downs and rough fields on the coast and inland; easily disturbed by day.

984(319). **C. incertana** (Treits.)
O. 6-7. Laid singly or in batches on a wide range of herbaceous plants.

L. 9-6. Mining at first; after hibernation in spun leaves, twisting those of tender shoots.
P. 6. In a folded leaf, or amongst detritus.
I. 6-7. Easily disturbed by day; sometimes rests on trunks and fences.

985(305). **Tortricodes alternella** ([D. & S.]) (*tortricella* (Hübn.))
O. 4. On *Quercus* spp., *Carpinus, Betula* spp., *Corylus, Crataegus, Prunus spinosa* or *Tilia*.
L. 5-6. In spun leaves, or under a turned-down leaf-edge.
P. 7-2. In a cocoon of earth and silk, on the ground or amongst detritus.
I. 2-3. The male flies in the sunshine; some adults emerge in January in mild spells.

986(304,304A). **Exapete congelatella** (Clerck) (*duratella* sensu Griffith)
O. 10-4. Laid in batches of two or three on twigs of *Ligustrum, Rhamnus, Crataegus, Erica, Calluna, Vaccinium* spp., *Potentilla* or *Rubus*.
L. 5-7. In spun shoots.
P. 6-10. Spun up in detritus on the ground.
I. 10-12. The male is easily beaten out by day, flies before dusk and comes to light; the brachypterous female rests on shoots of the foodplant after dark.

987(320). **Neosphaleroptera nubilana** (Hübn.)
O. On *Crataegus, Prunus* or, rarely, *Malus*.
L. 9-6. In spun shoots.
P. 6. In the larval habitation.
I. 7. Returns quickly to cover when disturbed. The male flies round the foodplants in sunshine and during the evening; the female towards dusk.

988(307). **Eana argentana** (Clerck)
O. Laid in batches on the foodplant.
L. *Recorded on grasses, herbaceous plants and mosses; not feeding till after hibernation.
P.
I. 6-7. Flies feebly, rather like *Crambus perlellus*, at dusk. Possibly extinct in Britain, but may have been overlooked in its mountain habitat.

989(306). **E. osseana** (Scop.)
O.
L. 5-6. Polyphagous on herbaceous plants. In a silken tube among the rootstocks and under stones.
P. 6-7. Spun up in, or near, the larval habitation.
I. 7-8;9 in the north. Readily disturbed by day.

990(311). **E. incanana** (Steph.)
O. On *Endymion non-scriptus* or *Leucanthemum vulgare*.
L. 4-5. In a silken web spun over the flowers.
P. 6. Spun up in ground debris.
I. 7. Both sexes fly late at night and come to light.

991(309). **E. penziana penziana** (Thunb. & Becklin) (*bellana* (Curt.))
O. 7. On *Festuca ovina*.
L. 7-6. In a silken gallery at the roots, feeding at night on the grass tips.
P. 5-6. In the larval habitation or a rock-crevice.
I. 6-7. Inactive by day; flies after dusk, and comes to light.

992(310). **E. penziana colquhounana** (Barr.)
O. 5-9. On *Plantago maritima* or *Armeria maritima*, the period of oviposition varying according to the locality.
L. 8-6. On the leaves, from a tube among the roots.
P. 5-7. In the larval habitation or a rock-crevice.
I. 5-9. Habits similar to those of no. 991, of which it is considered to be a subspecies.

993(288). **Aleimma loeflingiana** (Linn.)
O. On *Quercus* spp., *Carpinus* or *Acer* spp.
L. 5. In a rolled or folded leaf.
P. 6. In a dull white cocoon in the leaf, or on the ground.
I. 6-7. Flies freely at dusk; rests by day at base of trunks or on the ground.

994(289). **Tortrix viridana** (Linn.)
O. 7-4. Laid in small batches on the bark of branches of *Quercus* spp., or, occasionally, on other trees.
L. 4-6. In the buds; then in rolled or folded leaves.
P. 6. In a folded leaf of the foodplant or the foliage below.
I. 6-8. Easily disturbed by day.

995(326). **Spatalistis bifasciana** (Hübn.)
O. 5-6. On *Rhamnus, Frangula* or *Swida*.
L. 7-5. In the berries.
P. 5.
I. 5-6. Can occasionally be beaten out by day; flies rather high just before dusk.

996(324). **Croesia bergmanniana** (Linn.)
O. 7-4. On a stem of *Rosa* spp.
L. 4-5. In spun leaves and shoots.
P. 5-6. In a folded leaf-edge, or the larval habitation.
I. 6-7. Easily disturbed by day; flies from late afternoon onwards.

997(323). **C. forsskaleana** (Linn.)
O. 8-9. On *Acer campestre* or *A. pseudoplatanus*.
L. 9-6. First in spun leaves or flowers; then in a longitudinally rolled leaf.
P. 6-7. In a white silken cocoon spun in a folded leaf of the foodplant.
I. 7-8. Rests on leaves by day and is easily boxed.

998(328). **C. holmiana** (Linn.)
O. On *Crataegus, Rosa, Rubus, Malus, Pyrus, Prunus* or *Cydonia*.
L. 5-6. Between two leaves spun together at the margins.
P. 6-7. In the larval habitation, or a folded leaf.
I. 7-8. Easily disturbed by day; comes regularly to light.

999(338). **Acleris laterana** (Fabr.) (*latifasciana* (Haw., 1811))
O. 9-4. Laid singly at the base of a bud of *Crataegus, Prunus, Rosa, Rubus, Sorbus, Salix* or *Vaccinium*.
L. 5-6. In spun leaves and flowers.
P. 7-8. In a folded leaf or the larval habitation.
I. 8-9. Returns quickly to cover when disturbed; flies from dusk onwards.

1000(337). **A. comariana** (L. & Z.)
O. 10-4;7. Laid singly, generally on the petiole of a leaf of *Potentilla palustris, Fragaria* or *Geum rivale*.
L. 5-6;7-8. In folded or spun leaves.
P. 6-7;8-9. In the larval habitation.
I. 6-7;8-10. Flies in the afternoon and at dusk. Sometimes a pest of cultivated strawberries.

1001(333). **A. caledoniana** (Steph.)
O. On *Vaccinium myrtillus*; also on *V. vitis-idaea, Myrica, Alchemilla alpina, Rubus chamaemorus, Potentilla* or *Pinus contorta*.
L. 6-7. In spun leaves and shoots.
P. 6-7. Between spun leaves, or in the larval habitation.
I. 7-9. Flies in sunny or warm afternoons.

1002(345). **A. sparsana** ([D. & S.])
O. On *Fagus* or *Acer pseudoplatanus*.
L. 6-8. At first in a slight web on the underside of a leaf; later living in a chamber between spun leaves as a base from which to feed on the surrounding foliage.
P. 7-8. In the leaves, or in a slight cocoon on the ground.
I. 8-10. Can readily be disturbed by day; flies from dusk onwards.

1003(330). **A. rhombana** ([D. & S.]) (*contaminana* (Hübn.))
O. 9-4. Singly or in small batches on the bark of *Crataegus, Prunus, Malus, Sorbus, Pyrus* or *Rosa*.
L. 4-6. In the buds, then spun leaves, usually terminal, and flower-heads.
P. 6-7. In the larval habitation, or between spun leaves; pupal stage longer than the usual three weeks of related species.
I. 8-10. Easily disturbed; flies at dusk and comes to light.

1004(332). **A. aspersana** (Hübn.)
O. On *Filipendula, Potentilla, Fragaria vesca*, etc.
L. 5-6. In folded or spun leaves.
P. 6-7. In a folded leaf, or where the larva fed.
I. 7-8. Flies freely in the afternoon and evening and later comes to light.

1005(335). **A. ferrugana** ([D. & S.]) (*fissurana* (P. & M.))
O. 4-5;7-8. On *Quercus* spp. or *Salix caprea*.
L. 5-6;8. Between two leaves spun together.
P. 6-7;9. In spun leaves.
I. 7;9-4. Easily disturbed by day.

1006(334). A. **notana** (Don.) (*tripunctana* (Hübn.); *ferrugana* sensu Ford)

O. 4-5;7-8.	On *Betula* spp., *Alnus glutinosa* or *Myrica*.
L. 5-6;8-9.	In spun leaves bunched together.
P. 5-6;9-10.	Usually in the larval habitation.
I. 7;10-5.	Easily disturbed by day. Univoltine in Scotland.

1007(336). A. **shepherdana** (Steph.)

O.	On *Filipendula ulmaria*.
L. 5-6.	In spun shoots.
P. 6-7.	In the larval feeding place, a folded leaf-edge, or on the ground.
I. 8-9.	Not easily put up by day. Flies at night, coming readily to light.

1008(339). A. **schalleriana** (Linn.)

O. 4-5.	On *Viburnum lantana* or *V. opulus*.
L. 6-8.	In a pocket in a twisted leaf.
P. 8-9.	In the pocket.
I. 8-5.	Can be beaten out by day.

1009(341). A. **variegana** ([D. & S.])

O.	Laid singly, or in small batches, on *Rosa, Rubus, Crataegus, Prunus, Malus, Pyrus* or *Poterium sanguisorba*, usually on the midrib of a leaf.
L. 5-7.	Under a folded leaf-edge, or between spun leaves.
P. 7.	In the feeding place, or amongst dead leaves on the ground.
I. 7-9.	Easily disturbed by day; flies from dusk.

1010(340). A. **permutana** (Dup.)

O. 8-5.	Laid in small batches on leaves or stems of *Rosa pimpinellifolia* or *Prunus spinosa*.
L. 6-8.	In spun leaves and shoots.
P. 8.	In a white silken cocoon spun in leaves on the plant or on detritus beneath.
I. 8-9.	Reluctant to move by day; flies from dusk onwards.

1011(352). A. **boscana** (Fabr.)

O. 3-4;7-8.	Laid in small batches on *Ulmus* spp.
L. 5-6;8-9.	Between two leaves spun one above the other.
P. 6-7;9.	Between two leaves, or in a folded leaf.
I. 6-7;9-4.	May be disturbed from the elms in warm weather.

1012(351). A. **logiana** (Clerck)

O. 4-5.	Laid singly or in small batches on the twigs of *Betula* spp.
L. 6-7.	In spun leaves.
P. 7-8.	Between spun leaves.
I. 9-4.	Flies at night; may be found resting on birch-trunks after hibernation.

1013(344). A. **umbrana** (Hübn.)

O.	On *Sorbus, Prunus padus, Swida, Crataegus, Salix, Carpinus* or *Alnus glutinosa*.
L. 6-7.	
P. 7-8.	
I. 8-4.	Appears to rest high up, and is rarely seen by day; flies at night and visits ivy and sallow blossom.

1014(349). **A. hastiana** (Linn.)

O. 3-4;7. On the bark of twigs (first generation) or terminal leaves (second generation) of *Salix* spp., particularly *S. repens* and *S. aurita* but recorded also on *S. fragilis, Populus alba* and *Prunus spinosa*.

L. 4-6;7-8. In spun leaves, often in the young shoots, fastening them to the stem.

P. 6-7;8-10. In a flimsy cocoon in spun or folded leaves, usually away from the feeding place or on the ground.

I. 6-7;8-4. Seldom seen by day, but flies in the night; first brood comes to light more readily than the second. Bivoltine only as far north as Wigtonshire in Scotland, and in west and south-west Ireland.

1015(348). **A. cristana** ([D. & S.])

O. 3-4. Laid singly or in batches of 2 or 3 on the bark below a bud, particularly on *Prunus spinosa*; also on *Crataegus, Malus* or *Pyrus*.

L. 4-7. First in the opening buds, then in a folded leaf-edge, finally in spun leaves.

P. 6-8. In a flimsy cocoon in a folded leaf, or in debris on the ground.

I. 8-5. May be beaten from the foodplant from August to October or later, falling without flying in cold and wet weather.

1016(347). **A. hyemana** (Haw.)

O. 4-5. On *Calluna* or *Erica*; also *Pinus contorta* in Scotland.

L. 6-8. In a slight web on the top shoots, preferring the flowers.

P. 8-9. In an earthen cocoon; sometimes in the feeding place.

I. 9-4. Flies on sunny afternoons after hibernation.

1017(346). **A. lipsiana** ([D. & S.])

O. 4-5. In small batches on leaves of *Myrica, Vaccinium myrtillus* or *V. vitis-idaea*; also on *Betula* spp. or *Malus*.

L. 6-7. In a tube formed of spun leaves, longer (up to 7 cm) than that made by *A. maccana*.

P. 7-8. In the feeding place or on the ground.

I. 8-5.

1018(342). **A. rufana** ([D. & S.])

O. 4-5. On *Myrica*; also *Salix caprea, Populus alba*, or *Filipendula ulmaria*.

L. 6-8. In a neat tube of leaves spun round a stem.

P. 8. In a silken cocoon amongst dead leaves or in the earth.

I. 8-4. Flies from the evening onwards.

1019(331). **A. lorquiniana** (Dup.)

O. 7;?. On *Lythrum salicaria*.

L. 5-6;7-8. The first generation in the young shoots, the second in the flower spike.

P. 5-6;8-9. Amongst debris on the ground.

I. 6-7;9-10. Lethargic by day; flies at night, and comes readily to light.

1020(–). **A. abietana** (Hübn.) *Ent.Rec.* **79**:151-152(1967);**89**:53(1977)

O.

L. *6-7. In a slight web between the needles of *Pinus, Abies* or *Picea*.

P. *7-8. In the larval habitation.
I. 10. Probably overwinters. The early stages have not yet been observed in Britain.

1021(343). **A. maccana** (Treits.)
O. 5-6. On *Vaccinium* spp. or *Myrica*.
L. 6-7. In a tube of leaves of a shoot spun together; more twisted and untidy than that of *A. rufana* and shorter than that of *A. lipsiana*.
P. 7-8. In the larval habitation.
I. 8-5. Rarely flies by day, but does so freely at night when it comes to light and sugar.

1022(350). **A literana** (Linn.)
O. 4-5. Singly or in small numbers on branches or twigs of *Quercus* spp.
L. 5-6. Amongst spun leaves which it skeletonizes. May be dislodged by beating.
P. 6-7. Between spun leaves.
I. 8-5. Rests on lichen-covered boughs and may be disturbed by beating.

1023(329). **A. emargana** (Fabr.) (*caudana* (Fabr.))
O. On *Salix* spp., *Populus, Betula* spp., *Corylus* or *Alnus glutinosa*.
L. 5-6. In a turned-over leaf-edge, eating the tip, or in spun shoots.
P. 6-7. In the larval habitation, fallen leaves, or under moss on the tree-trunks.
I. 7-9. Easily disturbed by day.

OLETHREUTINAE

1024(494). **Celypha striana** ([D. & S.])
O. 7-8. On *Taraxacum*.
L. 9-5. First on the surface of the root, later mining into it.
P. 5-6. In a cocoon in or near the roots.
I. 6-8. The male flies by day in sunshine; both sexes fly from dusk onwards, coming to light.

1025(493). **C. rosaceana** (Schläg.) (*purpurana* (Haw.) nec. (Thunb.))
O. On *Sonchus arvensis, S. asper* or *Taraxacum*.
L. 5-6. In a silken tube on the roots.
P. 6. In a tough cocoon just beneath the surface of the soil.
I. 6-7. Hides by day and not easily disturbed. Both sexes fly late in the evening and come to light.

1026(492). **C. rufana** (Scop.)
O. 8. On *Tanacetum vulgare* or *Artemisia vulgaris*.
L. 9-5. Beneath a silk webbing on the rootstock and roots.
P. 5-6.
I. 6-7. Frequents bramble bushes growing on stony ground. Flies at midday in sunshine and again in late afternoon and evening.

1027(477). **C. woodiana** (Barr.)
O. 8-9. On *Viscum*.
L. 9-5. Mines the leaves, making large blotches when full-grown.
P. 6. In a cocoon under bark or lichen on the tree.
I. 7-8. Rests on trunks by day. Flies from dusk onwards, coming to light.

1028(506). **Olethreutes cespitana** (Hübn.)
O. 8-9. On *Thymus* spp., *Limonium* spp., *Armeria maritima*, *Trifolium* spp., etc.
L. 9-5. In a silken tube, usually low on the plant; overwinters when small.
P. 5-6.
I. 6-8. Active by day, especially in sunshine.

1029(498). **O. rivulana** (Scop.)
O. On *Filipendula ulmaria, Genista* spp., *Orchis* spp., etc.
L. 5-6. On the flowers and terminal shoots, often causing them to wilt.
P. 6. In a flimsy cocoon in the larval habitation or amongst leaves of the foodplant.
I. 7-8. Active in sunshine by day; flies freely from late evening onwards, coming to light.

1030(509). **O. aurofasciana** (Haw.) (*latifasciana* (Haw., 1812, nec 1811))
O.
L. 4-5. In a silken gallery under moss growing on old or rotting bark of trees.
P. 5-6. In the larval habitation.
I. 6-7. Rests on trunks by day, when it is easily disturbed. Flies during late afternoon and early evening.

1031(491). **O. mygindiana** ([D. & S.]) (*mygindana* missp.)
O. 5-6. Laid in small batches on the lower leaves of *Vaccinium vitis-idaea*, or, less often, on *Arctostaphylos* or *Myrica*.
L. 6-5. In an untidy and conspicuous spinning at the tip of a shoot.
P. 5-6. In the larval habitation.
I. 5-6. Flies from late afternoon onwards; occasionally earlier in sunshine.

1032(495). **O. arbutella** (Linn.)
O. 7. On *Arctostaphylos*.
L. 8-5. In spun terminal leaves or between two leaves spun together.
P. 5-6. In the larval habitation or between freshly spun leaves.
I. 5-6. Flies in afternoon sunshine over the foodplant.

1033(501). **O. metallicana** (Hübn.)
O. On *Vaccinium myrtillus*.
L. 5-6. In spun leaves.
P. 6-7.
I. 6-7. Rests by day amongst the foodplant. Flies from late evening to dusk.

1034(499). **O. schulziana** (Fabr.)
O. 7-8.
L. *9-5. In a silken tube low down on *Vaccinium oxycoccus* or *Calluna*.

P. 5-6.
I. 6-8. Readily disturbed by day. Flies in the evening sunshine.

1035(500). **O. palustrana** (L. & Z.)
O.
L. 5-6. In a tubular silken spinning in mosses.
P. 5-6.
I. 5-7. Flies mainly in the evening and at dusk, but may become active earlier on sunny afternoons.

1036(502). **O. olivana** (Treits.) (*micana* sensu Hübn.)
O.
L. *5-6. Amongst herbaceous plants and mosses.
P. 6-7.
I. 7-8. The male rests openly on herbage and is easily disturbed in sunshine. Both sexes fly towards sunset.

1037(504). **O. lacunana** ([D. & S.])
O. 7-8. Laid in batches of two or three on the upper surface of a leaf along a vein, hatching in about 10 days.
L. 8-5. In spun leaves, shoots or flowers of most species of herbaceous plant; occasionally recorded on trees and shrubs.
P. 5-7. In a cocoon spun amongst leaf-litter.
I. 5-8. Readily disturbed by day and active at dusk.

1038(503). **O. obsoletana** (Zett.) (*irriguana* (Herr.-Schäff.))
O.
L.
P.
I. 7. Hides by day, but readily disturbed in sunny weather. Flies from dawn until after sunrise.

1039(505). **O. doubledayana** (Barr.)
O.
L.
P.
I. 7-8. Sluggish, but may be disturbed from dense vegetation on sunny afternoons. Flies from shortly before sunset till dusk.

1040(508). **O. bifasciana** (Haw.)
O.
L. 5-6. In a loose spinning amongst the shoots and male blossoms of *Pinus sylvestris* and *P. pinaster*.
P. 6. In a flimsy cocoon in the larval habitation.
I. 6-7. Readily beaten from pines by day. Begins to fly before dusk and later comes to light.

1041(497). **O. arcuella** (Clerck)
O. 7-9.
L. 9-4. On decaying leaves on the ground.
P. 4-5. In a cocoon amongst leaf-litter.
I. 5-8. Flies by day and basks in sunshine on the leaves of bushes or bracken.

1042(490). **Pristerognatha penthinana** (Guen.)
O. 7. On *Impatiens noli-tangere*.
L. 8-5. In the lower portion of the stem, burrowing down to the root.

P. 5-6. In the larval habitation.
I. 5-6.

1043(486). **Hedya pruniana** (Hübn.)
O. On *Prunus spinosa, P. domestica* or *P. avium*; less frequently on *Malus, Pyrus, Crataegus* or *Corylus*.
L. 4-5. In spun shoots and leaves.
P. 5-6. In the larval habitation or freshly spun leaves.
I. 5-7. Readily disturbed by day. Flies from late evening onwards.

1044(484). **H. nubiferana** (Haw.) (*variegana* Hübn.)
O. 7-8. Laid singly or in small batches on the underside of a leaf of *Crataegus, Prunus* or, less often, other trees.
L. 8-5. In spun shoots, leaves and flowers.
P. 5-6. In the larval habitation or freshly spun leaves.
I. 6-7. Readily disturbed by day. Flies from late evening onwards.

1045(485). **H. ochroleucana** (Fröl.)
O. On *Rosa* spp. or, occasionally, *Malus*.
L. 4-6. In a shoot spun into a large bunch.
P. 5-6. In the larval habitation.
I. 6-7. Readily disturbed by day. Flies from dusk onwards, coming to light.

1046(489). **H. atropunctana** (Zett.) (*dimidiana* (Sodof.) nec (Fröl.))
O. 7. On *Myrica, Salix* spp. or *Betula* spp.; occasionally on other trees.
L. 8-10. On *Myrica* in leaves spun together so as to form a hollow space about the size of a large acorn; on *Salix* and *Betula* spinning successive leaves into a pod.
P. 11-4. In a cocoon amongst leaf-litter.
I. 5-6. Rests low down amongst herbage by day; flies during the evening. In the south a second generation may occur in August.

1047(476). **H. salicella** (Linn.)
O. On *Salix alba*; less frequently on other species of *Salix, Populus tremula* or *P. nigra*.
L. 5-6. In a spun shoot or a folded leaf.
P. 6. In the larval habitation.
I. 6-8. Rests on trunks or the surface of a leaf, resembling a bird-dropping. Flies freely towards dusk and later comes to light.

1048(507). **Orthotaenia undulana** ([D. & S.]) (*urticana* sensu Hübn.)
O.
L. 4-6. In spun leaves of *Lonicera* spp., *Vaccinium* spp., *Myrica, Betula* spp. and a wide range of other trees, bushes and herbaceous plants.
P. 5-6. In the larval habitation or a freshly spun leaf.
I. 5-7. Readily disturbed by day. Flies towards sunset and later comes to light.

1049(510). **Pseudosciaphila branderiana** (Linn.)
O. On *Populus tremula*.
L. 5. In a folded or rolled leaf, or between two spun leaves.
P. 6. In the larval habitation or a freshly folded leaf.

I. 6-8. Rests high up by day. At dusk flies erratically round the aspens and later comes to light.

1050(478). **Apotomis semifasciana** (Haw.)
O. On *Salix* spp.
L. 5-6, On catkins and later in spun leaves.
P. 6-7. In a cocoon between spun leaves.
I. 7. Seldom seen by day. Flies at dusk and later comes to light and sugar.

1051(–). **A. infida** (Heinrich) *Ent.Gaz.* **14**:39(1963)
O.
L. *Has been recorded on *Salix*.
P.
I. 7. Taken in Perthshire in 1919.

1052(479). **A. lineana** ([D. & S.]) (*hartmanniana* (Linn.) nec (Clerck))
O. On *Salix alba* or *S. fragilis*.
L. 5. Between spun leaves.
P. 6.
I. 6-8. Occasionally found resting on trunks, but flies off at the slightest disturbance.

1053(480). **A. turbidana** (Hübn.) (*corticana* sensu Hübn., 1799)
O. On *Betula* spp.
L. 5. In spun leaves.
P. 6. In the larval habitation or freshly spun leaves.
I. 6-7. Rests on trunks and leaves by day. Flies towards dusk and later comes to light.

1054(481). **A. betuletana** (Haw.)
O. On *Betula* spp.
L. 5-6. Amongst spun leaves, rolling and twisting together two or three leaves.
P. 6-7. In the larval habitation, in a spun leaf, or amongst leaf-litter.
I. 7-9. May be disturbed by day. Flies towards dusk and later comes to light and sugar.

1055(482). **A. capreana** (Hübn.)
O. On *Salix caprea*.
L. 4-5. In tightly spun leaves.
P. 5-6. In the larval habitation or amongst leaf-litter.
I. 6-8. Flutters to the ground if disturbed. Flies towards dusk and later comes to light.

1056(483). **A. sororculana** (Zett.)
O. 6-7. Laid singly on either surface of a leaf of *Betula* spp.
L. 7-9. Between leaves spun flatly together.
P. 9-4. In the larval habitation or spun up between 2 leaves, which later fall to the ground.
I. 5-7. Rests high up by day, fluttering to the ground if dislodged. Flies towards dusk, later coming to light.

1057(488). **A. sauciana sauciana** (Fröl.)
O. On *Vaccinium myrtillus*.
L. 4-6. In a loosely spun terminal shoot.

P. 6. In the larval habitation.
I. 6-7, a month later in the north of its range. Rests on leaves of the foodplant and flies in the afternoon and evening. The southern subspecies.

1058(487). **A. sauciana grevillana** (Curt.)
O. On either surface of a leaf of *Vaccinium myrtillus* or *Arctostaphylos*.
L. 6 or 6-4. In a spun terminal shoot. Overwintering larvae do so in a silken hibernaculum spun on a leaf.
P. 4 or 6. In the larval habitation. June pupae will produce adults in the same year.
I. 5-8. Flies erratically over its foodplant on sunny afternoons. The northern subspecies.

1059(470). **Endothenia gentianaeana** (Hübn.) (*gentianana* (Hübn.))
O. 8. On *Dipsacus fullonum*.
L. 9-5. On the pith of a seedhead. Overwinters full-fed in the feeding-place.
P. 6. In the larval habitation, in a cocoon of silk and particles of pith.
I. 7. Occasionally seen resting, newly emerged, on teasel-heads.

1060(469). **E. oblongana** (Haw.) (*sellana* (Fröl.))
O. 8.
L. 9-5. Possibly on *Centaurea nigra*.
P. 5-6.
I. 6-7. Rests low down and is readily disturbed. Flies during the evening.

1061(468). **E. marginana** (Haw.) (*oblongana* sensu Ford)
O. 8-9. On *Betonica, Galeopsis* spp., *Pedicularis* or *Rhinanthus*.
L. 9-6. In the seeds, living externally in a silken gallery or internally, according to the foodplant.
P. 5-7.
I. 6-8. Readily disturbed by day. Flies low down during the evening.

1062(473). **E. pullana** (Haw.) (*fuligana* sensu auctt., nec Haw.)
O. On *Stachys palustris*.
L. 9-5. In the stem and rootstock.
P. 5-6. In the larval habitation, usually just below a node in the stem.
I. 6-7. Flies in sunshine on warm afternoons.

1063(472). **E. ustulana** (Haw.) (*carbonana* (Doubl.))
O. On *Ajuga reptans*.
L. 9-5. Overwinters in an early instar and feeds up in the spring, first in the rootstock, later in the leaf-stalks.
P. 5-6. In the larval habitation or amongst leaves of the foodplant.
I. 6-7. Flies actively in the afternoon, especially in sunshine.

1064(471). **E. nigricostana** (Haw.)
O. 6. On the flower-head or upper part of the flower-stalk of *Stachys sylvatica*.
L. 7-4. Mines down the stem into the root where it overwinters, returning upwards in the spring.

P. 5-6. In the upper portion of the stem in which the larva fed. The stems may be collected from March onwards.
I. 6. Rests openly on the leaves in the evening and flies towards dusk, later coming to light.

1065(474). **E. ericetana** (H. & W.)
O. 7-8. On *Stachys palustris* or, possibly, *S. sylvatica* or *Mentha arvensis*.
L. 8-5. In the lower stem and roots.
P. 5-6. In the larval habitation, or spun up in the soil or amongst leaf-litter.
I. 6-8. Rests low down during the day. Flies lethargically in the evening and later comes to light.

1066(475). **E. quadrimaculana** (Haw.) (*antiquana* (Hübn.))
O. 7. On *Stachys palustris*; occasionally on *Mentha spicata*.
L. 8-5. In a silk-lined gallery in the roots and underground stem.
P. 5-6. In a cocoon spun in the soil or amongst leaf-litter.
I. 6-7. Hides low down by day. Flies in the evening, and later comes to light.

1067(466). **Lobesia occidentis** Falkovitsh (*euphorbiana* sensu auctt.)
O. 6-7;8. On *Euphorbia amygdaloides* or *E. paralias*.
L. 7;8-9. In spun terminal shoots causing discoloration, preferring young plants not due to flower till the next year.
P. 7;10-5. In a dense cocoon between two leaves, in leaf-litter or in the larval habitation.
I. 6;8. Flies towards dusk and sometimes earlier in sunshine.

1068(467). **L. reliquana** (Hübn.) (*permixtana* (Hübn.))
O. 6. On either side of a leaf of *Quercus* spp., *Prunus spinosa* or *Betula* spp.
L. 7-4. In terminal shoots or spun leaves. Overwinters full-fed from September in a cocoon spun in a dead leaf or on the ground.
P. 4. In the cocoon.
I. 5-6. Rests by day amongst foliage. Flies in the early morning and again in the late afternoon and evening in the more open parts of woods.

1069(–). **L. botrana** ([D. & S.]) *Ent.Gaz.* 28:21-22(1977)
O. *On the fruit or stem of various shrubs and climbing plants, including *Lonicera* spp., *Clematis, Viburnum, Ligustrum, Ribes, Hedera*.
L. *5-10. In two or more generations in spun flowers or on the fruit.
P. *In the larval habitation or in bark.
I. 7. A pest of vineyards on the Continent.

1070(465). **L. abscisana** (Doubl.) (*fuligana* sensu Haw.)
O. On *Cirsium arvense*.
L. 6;8-9. In spun shoots.
P. 5;7. In a cocoon spun amongst surface litter.
I. 5;7-8. Rests concealed by day; flies at dusk, occasionally earlier, and later comes to light.

1071(464). **L. littoralis** (H. & W.)
O. 7;?. On *Armeria maritima*.

L. 4-5;8. Spring larvae in the flower-heads and shoots, and on leaves; autumn larvae on the seeds in dead flower-heads, causing distortion.
P. 5-6;8-9. In a cocoon within the flower-head or spun amongst the roots.
I. 7;9-10. Flies in afternoon sunshine and later sometimes comes to light. Occurs with its foodplant both on salt-marshes and in gardens.

1072(462). **Bactra furfurana** (Haw.)
O. On *Scirpus lacustris* or *Juncus subuliflorus*.
L. 4-5. In a stem, exuding frass and causing the stem to turn brown. Moves from stem to stem.
P. 5-6. In a slight cocoon in the stem in which the larva last fed.
I. 6-7. Rests low down. Flies in the afternoon and, when the weather is warm, continues till dusk and later comes to light. The female is restless when boxed.

1073(461). **B. lancealana** (Hubn.) (*lanceolana* (Hübn.))
O. On *Juncus subuliflorus, Scirpus lacustris, S. cespitosus* or *Cyperus longus*.
L. 4-5;7-8. In the stem.
P. 5;8. In a slight cocoon in the stem in which the larva has fed.
I. 5-8;8-10. Two overlapping generations. Readily disturbed by day; is also attracted to light.

1074(463). **B. robustana** (Christoph) (*scirpicolana* P. & M.)
O. On *Scirpus maritimus*.
L. 8-6. In the stem.
P. 6. In a strong white cocoon in the larval habitation. A stem which contains a pupa shows signs of decay and is usually black in the region of the cocoon.
I. 6-7. Lethargic by day but may be flushed with a bee-smoker. Flies in the evening and later comes to light.

1075(512). **Eudemis profundana** ([D. & S.])
O. On *Quercus* spp.
L. 6. In a leaf rolled round the midrib.
P. 7. Spun up in the larval habitation or leaf-litter.
I. 7-8. Rests on branches by day, and is easily dislodged. Flies rather high up towards sunset and later comes to light.

1076(511). **E. porphyrana** (Hübn.) (*pomedaxana* P. & M.)
O. On *Malus*, especially old crab-apple trees; possibly also on *Quercus* spp.
L. 5-6. In a leaf rolled round the midrib; sometimes in two leaves rolled together.
P. 6-7. In the larval habitation or a folded leaf.
I. 7-8. Rests by day on the foliage or trunks and may be obtained by beating. Flies actively from evening onwards and later comes to light.

1077(383). **Ancylis achatana** ([D. & S.])
O. On *Crataegus, Prunus spinosa* or, occasionally *Cotoneaster* spp.

L. 5. Feeds on the surrounding foliage from a tubular shelter of spun leaves.
P. 5-6. In loosely spun leaves of the foodplant. Obtained by beating.
I. 6-7. Rests on the foliage by day. Flies from dusk onwards, later coming to light.

1078(373). **A. comptana** (Fröl.)
O. 5-6;8. Laid singly, usually on the underside, on a leaf of *Poterium sanguisorba, Potentilla* spp., *Thymus* or *Fragaria*.
L. 7;9-10. When young on the underside of a leaf, under a slight web; later in a folded leaf, feeding on the upper surface. Some second generation larvae spin their cocoons in the autumn; others feed slowly through the winter and spin up in the spring.
P. 4-5;end 7. In a cocoon spun in a dead leaf.
I. 4-6;7-8. Readily disturbed by day. Flies in afternoon sunshine and later comes to light.

1079(374). **A. unguicella** (Linn.)
O. 6-early 7. Laid singly on the upper surface of leaves of *Erica cinerea* or, occasionally, *Calluna*.
L. 7-4. Feeds on the leaves. Overwinters in a cocoon spun amongst leaf-litter.
P. 4. In the cocoon.
I. 5-early 7. Flies in the morning and from late afternoon till sunset. Occasionally comes to light.

1080(375). **A. uncella** ([D. & S.]) (*uncana* (Hübn.))
O. 5-6. On the upper surface of a leaf or a stem of *Erica* spp. or *Betula* spp.
L. 7-4. In leaves spun together to form a pocket. Overwinters in a cocoon spun up in a dead leaf.
P. 4. In the cocoon.
I. 5-6. Sometimes a small second generation in August. Rests in heather by day. Flies from sunset onwards and comes to light.

1081(377). **A. geminana** (Don.) (*biarcuana* (Steph.))
O. 6-8. On *Salix* spp. other than *S. repens*.
L. 7-4. In a leaf spun into a pod. Hibernation similar to that of *A. diminutana*.
P. 4-5. In the cocoon.
I. 5-8. Flies in the afternoon.

1082(376). **A. diminutana** (Haw.)
O. 6-8. On *Salix* spp. other than *S. repens*.
L. 7-4. In a leaf spun into a pod. When full-fed the larva spins a cocoon amongst leaf-litter in which it overwinters.
P. 4-5. In the cocoon.
I. 5-8. Flies at dusk, later occasionally coming to light.

1083(378). **A. subarcuana** (Dougl.) (*inornatana* (Herr.-Schäff.))
O. 6-8. On *Salix repens*.
L. 7-9. In a leaf spun into a pod.
P. 10 or 4-5. In captivity pupates in the autumn, producing adults in October.

I. 5-8. Flies in the afternoon. Nos 1081-1083 may be biological races of a single species.

1084(380). **A. mitterbacheriana** ([D. & S.])

O. 6-7. On *Quercus* spp. or *Fagus*.
L. 7-4. In a leaf spun into a pod. Overwinters full-fed in the last feeding place.
P. 4-5. In the pod where the larva overwintered, or in leaf-litter.
I. 5-6. Easily disturbed by day. Flies in the evening and later comes to light.

1085(381). **A. upupana** (Treits.)

O. 6-7. On *Betula* spp. or *Ulmus* spp.
L. 7-4. In a leaf spun into a pod, or between two leaves neatly spun together to form a pocket. It partly skeletonizes the leaves. Overwinters full-fed.
P. 4-5. In the larval habitation which has fallen to the ground.
I. 5-6. Rests high up and, if dislodged, flutters to the ground. Flies round the tree-tops in the afternoon.

1086(372). **A. obtusana** (Haw.)

O. On *Rhamnus* or *Frangula*.
L. 9-4. Between spun leaves. Overwinters full-fed.
P. 4-5. In the larval habitation.
I. End 5-early 7. Rests on the foliage. Flies in sunshine during late afternoon.

1087(382). **A. laetana** (Fabr.)

O. 6-7. On *Populus tremula*; occasionally on *P. nigra*.
L. 7-4. In a pocket in a folded leaf or between two leaves neatly spun together. Overwinters full-fed in the last spinning, which falls to the ground.
P. 4-5. In the larval habitation.
I. 5-6. Rests on foliage and trunks. Flies at sunset, later coming to light.

1088(379). **A. tineana** (Hübn.)

O. *5-6;8-9. On *Crataegus, Prunus spinosa, Betula* spp., *Populus tremula, Malus* or *Pyrus*.
L. *6-7;9-4. Between spun terminal leaves. The second generation overwinters full-fed.
P. *4-5;6-7. In a white cocoon in the larval habitation.
I. 4-5;7-8. May be disturbed in the afternoon.

1089(367). **A. unculana** (Haw.) (*derasana* (Hübn.))

O. 7-8. On *Rhamnus* or, less often, *Frangula*.
L. 8-5. In a leaf spun into a pod, preferring a lower leaf or one inside the bush. When full-fed, it spins a cocoon in the pod, in which it overwinters.
P. 5-6. In the pod.
I. 6-7. Easily disturbed by day. Flies from early evening onwards, later coming to light.

1090(368). **A. badiana** ([D. & S.]) (*lundana* (Fabr.))

O. 6;8.	On *Lathyrus* spp., *Vicia* spp. or *Trifolium* spp.
L. 6;9-4.	Folds a leaf or spins two together, according to the food-plant, and feeds therein. Second generation overwinters full-fed.
P. 4;7.	In the larval habitation or a leaf on the ground.
I. End 4-mid 6;7-8.	Flies at sunrise and in the evening, and occasionally comes to light.

1091(369). **A. paludana** (Barr.)

O. 6;8-9.	On *Lathyrus palustris*.
L. 6;9-4.	First mining, then between two leaflets neatly spun into a pod.
P. 4;7.	In a silken cocoon amongst plant-litter.
I. End 5-6; end 7-8.	Rests low down by day. Flies at dusk.

1092(370). **A. myrtillana** (Treits.)

O. 7.	Laid singly or in small batches, usually on the underside, on a leaf of *Vaccinium myrtillus* or *V. uliginosum*.
L. 7-4.	In a leaf spun into a pod. Overwinters full-fed.
P. 4.	In a folded leaf amongst leaf-litter on the ground.
I. End 5-early 7.	Active in morning sunshine and again towards sunset.

1093(371). **A. apicella** ([D. & S.]) (*siculana* (Hübn.))

O.	On *Frangula* or *Rhamnus*.
L. 6; end 7-4.	In a folded leaf; later in two leaves spun flatly together, blanching them in irregular patches. Summer generation overwinters full-fed from August in its cocoon.
P. 4;early 7.	In cocoon spun in the larval habitation or in leaf-litter.
I. 5-6;7-8.	Flies at dusk and comes to light.

1094(410). **Epinotia pygmaeana** (Hübn.)

O.	On *Picea* spp. or *Abies alba*.
L. 6-7.	At first mining a young needle; later amongst spun needles.
P. 8-4.	In a cocoon spun in leaf-litter or in the soil.
I. 4-early 6.	The male flies round the tops of the host trees about mid-day; the female, if disturbed, usually drops to the ground.

1095(411). **E. subsequana** (Haw.) *Ent.Rec.* **90**:67-68(1978)

O.	Laid in small batches on the upper surface or near the base of a needle of *Picea alba*, or, perhaps principally, of *Abies grandis*.
L. 6-7.	At first mining the leaves; later in spun terminal shoots.
P. 8-4.	In a cocoon spun in leaf-litter, or in the soil.
I. 4-5.	Flies in sunshine high up round the host trees from noon till late evening.

1096(440). **E. subocellana** (Don.)

O.	On rough-leaved species of *Salix*.
L. 8-4.	Between two leaves spun flatly together, feeding on the underside of the upper leaf. Overwinters in a cocoon spun at ground level.
P. 4-5.	In the cocoon.
I. End 5-early 7.	Flies towards dusk and later comes to light.

1097(443). E. **bilunana** (Haw.)

O. On *Betula* spp.
L. 9-4. In a catkin, distorting it.
P. 5. In the catkin in which it fed or amongst leaf-litter.
I. 6-7. Rests on trunks in daytime; flies towards dusk and later comes to light.

1098(442). E. **ramella** (Linn.)

O. On *Betula* spp.
L. 4-5. In a twig or catkin.
P. 6.
I. 7-9. Rests on trunks by day; flies towards dusk and later comes to light.

1099(447). E. **demarniana** (Fisch. v. Rösl.)

O. On *Betula* spp., *Alnus glutinosa* or *Salix caprea*.
L. 9-5. In a catkin.
P. 5-6. In a cocoon amongst leaf-litter.
I. 6-7. Rests high up in the daytime. Best obtained by kicking the trees in the morning when the moth flutters to the ground. It flies high about the trees in the evening and later comes to light.

1100(449). E. **immundana** (Fisch.v.Rösl.)

O. On *Alnus* spp. or *Betula* spp.
L. 9-3;7. In catkins and buds; the smaller second generation in rolled leaves.
P. 3-4;7-8. In a silken cocoon amongst leaf-litter; in earth or under moss.
I. 4-6;8-9. Flies towards dusk and later comes to light; rests on the trunks by day.

1101(448). E. **tetraquetrana** (Haw.)

O. On *Betula* spp. or *Alnus glutinosa*.
L. 7-10. When young, boring into a twig and causing a slight swelling; in the last instar in the turned down edge of a leaf.
P. 10-4. In a cocoon amongst leaf-litter.
I. 4-5. Flies actively in the evening and later comes to light.

1102(444). E. **nisella** (Clerck)

O. On *Salix*, especially the rough-leaved species, or on *Populus* spp.
L. 4-6. In the catkins or between two leaves spun flatly together.
P. 6-7. In a flimsy cocoon in the larval habitation or leaf-litter on the ground.
I. 7-8. Rests on the trunks by day; flies towards dusk and later comes to light.

1103(445). E. **cinereana** (Haw.) (*cinerana* missp.)

O. On *Populus tremula*.
L. 4-6. Life history similar to that of preceding species.
P. 6-7.
I. 7-8. May be a form of the preceding species.

1104(441). E. **tenerana** ([D. & S.]) (*penkleriana* sensu auctt.)

O. 10. Laid singly or in a small batch on a bud of *Corylus* or *Alnus glutinosa*.
L. 10-5. At first in the catkins, causing slight distortion; later in developing leaf-buds.
P. 6. In a flimsy cocoon in leaf-litter or under moss.
I. 7-9. Flies towards dusk and later occasionally comes to light.

1105(450). E. **nigricana** (Herr.-Schäff.)

O. 7-9. Laid singly on a leaf of *Abies alba*.
L. 9-4. In the buds, preferring the terminal buds of side shoots.
P. 5-6. In a fragile cocoon amongst leaves of the foodplant or in detritus on the ground.
I. End 6-7. Flies round the higher branches in afternoon sunshine.

1106(453). E. **nemorivaga** (Tengst.)

O. On *Arctostaphylos*.
L. 9-5. Mines the leaves, turning the upper half from red to black. In the spring it spins young leaves of a shoot loosely together, and moves from leaf to leaf, making inflated mines.
P. 5-6. In the larval habitation, usually in a mined leaf.
I. 6-7. Flies in afternoon sunshine.

1107(452). E. **tedella** (Clerck)

O. On *Picea abies*.
L. 8-4. First mines terminal leaves of side shoots, later feeds in a spinning which bunches the leaves and discolours them; only those which are still greenish white contain larvae. In mild winters full-fed by February.
P. 2-4. In a silken cocoon amongst leaf-litter.
I. 5-6. Readily disturbed; flies in afternoon sunshine and later comes to light.

1108(451). E. **fraternana** (Haw.) (*proximana* (Herr.-Schäff.)) *Ent.Rec.* **90**: 68(1978)

O. 7. Laid singly on a leaf of *Abies alba* or *A. grandis*, usually near the tip; occasionally on other species of *Abies*.
L. 8-4. Mines the leaves from the centre, first to the tip, then to the base; later in a neat spinning sometimes resembling the "frame for a game of noughts and crosses", within which it continues to feed by mining. When full-fed, it spins a cocoon amongst leaf-litter, in which it overwinters.
P. 4-5. In the cocoon.
I. 5-7. Flies high round the foodplant in afternoon sunshine and may be disturbed by tapping the branches.

1109(408). E. **signatana** (Dougl.)

O. On *Prunus* spp., *Malus sylvestris*, or **Crataegus*.
L. 5. In a folded leaf or spun shoot.
P. 5-6.
I. 6-7. Flies at dusk and later comes to light.

1110(409). E. **nanana** (Treits.)

O. On *Picea abies* or *P. stichensis*.
L. 9-5. Mines a leaf, overwintering half-fed in its mine; it moves to

a new needle via a silken tube along which it retreats to deposit frass in the previously mined leaf, a habit characterising the species.

P. 6. In a white cocoon in the larval tube or amongst leaf-litter.

I. 6-8. Readily disturbed from the foodplant; flies freely towards dusk and later comes to light.

1111(400). E. **rubiginosana** (Herr.-Schäff.)

O. On *Pinus sylvestris* or *P. pinea*.

L. 9-10. Inhabits a tubular spinning amongst the leaves, from which it emerges to feed on the adjacent needles.

P. 4-5. In a cocoon amongst leaf-litter or in the soil.

I. 6. Flies about the tops of the trees in afternoon sunshine onwards till sunset, and later comes to light.

1112(398). E. **cruciana** (Linn.)

O. On *Salix* spp., especially *S. cinerea, S. repens* or *S. aurita*.

L. 4-early 6. In an unopened leaf-bud or in spun leaves, especially those of a terminal shoot.

P. 5-6. In a flimsy cocoon amongst leaf-litter, usually in a fold of a dead leaf.

I. 6-early 8. Flies rather high round the foodplant from late afternoon onwards, later coming to light. *f. augustana* Hübn. is associated with *S. repens*.

1113(397). E. **mercuriana** (Fröl.)

O. On *Calluna, Vaccinium myrtillus, Dryas* or possibly other herbaceous plants.

L. 9-6. In spun leaves.

P. 6-7. In a cocoon in the soil.

I. 7-9. Flies at dusk and also, sometimes, in afternoon sunshine.

1114(446). E. **crenana** (Hübn.)

O. On *Salix cinerea, S. aurita* and possibly other *Salix* spp.

L. 5-7. In a spun terminal shoot.

P. 7-8. In a spun leaf.

I. 8-4. Flies in sunshine round the foodplant.

1115(407). E. **abbreviana** (Fabr.) (*trimaculana* (Don.))

O. On *Ulmus* spp. or, less often, *Acer campestre*.

L. 4-early 6. At first in an unopened leaf-bud; this feeding is conspicuous later as an arc of holes near the leaf-base. Later in spun leaves, often spinning one above another.

P. 6. In a cocoon amongst leaf-litter.

I. 6-7. Rests by day on the trunks and foliage; flies towards dusk.

1116(454). E. **stroemiana** (Fabr.) (*similana* (Hübn.) nec ([D. & S.]))

O. On *Betula* spp., favouring small trees.

L. 6-7. In a folded leaf or between spun leaves, usually at the end of a shoot.

P. 8. In a flimsy cocoon in the larval habitation or in the soil.

I. 8-9. Flies at dusk.

1117(455). E. **maculana** (Fabr.) (*ophthalmicana* (Hübn.))

O. On *Populus tremula* or, occasionally other *Populus* spp.

L. 6. In a rolled leaf, when nearly full-fed leaving it at night to feed in the adjacent foliage.

P. 7. In a flimsy cocoon in earth or amongst leaf-litter.
I. 8-10. Rests on the trunks; if dislodged by tapping, it settles again low down. Flies towards dusk.

1118(459). **E. sordidana** (Hübn.)
O. On *Alnus glutinosa*.
L. 5-6. At first in a rolled or puckered leaf; later making a pod of a leaf, usually a terminal leaf.
P. 7. In a tough cocoon in the soil.
I. 8-10. Readily flushed by day; flies low down towards dusk and later comes to light.

1119(458). **E. caprana** (Fabr.) (*semifuscana* (Steph.))
O. On *Salix* spp. or *Myrica*.
L. 4-6. In spun terminal leaves.
P. 7. Spun in leaf-litter on the ground, or in an earthen cocoon.
I. End 7-8, sometimes 9 or early 10. Flies at dusk and later comes to light.

1120(457). **E. brunnichana** (Linn.)
O. On *Betula* spp., *Corylus* or *Salix* spp.
L. 5-6. In a leaf rolled transversely from apex to petiole, eating the whole thickness of the leaf at the tip.
P. 6-7. In earth.
I. 7-8. Flies at dusk and later comes to light.

1121(456). **E. solandriana** (Linn.)
O. On *Betula* spp., *Corylus* or *Salix caprea*.
L. 5-6. In a leaf rolled longitudinally, along the axis of the midrib.
P. 6-7. In earth.
I. 7-8. Flies at dusk and later comes to light and, occasionally, sugar.

1122(389). **Crocidosema plebejana** Zell. (*plebeiana* missp.)
O.
L. 8-10. In the seeds of *Lavatera arborea, Althaea* spp. and *Malva* spp. Two holes in a capsule indicate that the larva has left.
P. 7-8. In a flimsy cocoon amongst leaf-litter.
I. 7-10. Flies from late afternoon onwards and comes to light.

1123(412). **Rhopobota ustomaculana** (Curt.)
O. 6-7. Laid in small batches on a leaf or stem of *Vaccinium vitis-idaea* or, possibly, *V. myrtillus*.
L. 7-5. In spun leaves, eating the upper epidermis and the parenchyma and leaving the lower epidermis intact; distinguished from the spinning of *Olethreutes mygindiana* in that the white silken spinning is scarcely visible.
P. 5-6. In the larval habitation.
I. 6-7. Flies in the afternoon sunshine and in the evening.

1124(355,356). **R. naevana** (Hübn.) (*unipunctana* (Haw.); *geminana* (Steph.))
O. 9-5. Laid singly on a twig of the foodplant.
L. 5-6. In spun leaves or, occasionally, flowers. Polyphagous on trees and shrubs, especially *Ilex* and *Prunus spinosa*.
P. 6-7. In the larval habitation or in a cocoon amongst detritus on the ground.

I. End 6- early 9. Flies towards dusk and later comes to light. A heathland race feeding on *Vaccinium myrtillus* has been regarded as a distinct species ((356) *R. geminana).*

1125(357). **Acroclita subsequana** (Herr.-Schäff.) (*consequana* (Herr.-Schaff.))
O. 5-6;8-10. Laid singly on a leaf of *Euphorbia paralias* or *E. portlandica.*
L. 6;9-3. On the leaves and seeds from a silken gallery, drawing the leaves against the stem. August larvae overwinter in their cocoons.
P. 3-4;7. In a cocoon, usually spun in the larval habitation.
I. 4-6;8-10. Flies from evening onwards and visits sugar.

1126(414). **Griselda stagnana** ([D. & S.]) (*fractifasciana* (Haw.))
O. On *Scabiosa* or *Succisa.*
L. 6-7;9-3. On the flowers and seeds in July; on the young leaves at the base of the plant in September, then overwintering full-fed.
P. 3-4;7-8. In a cocoon in the soil.
I. 4-6;8-9. Flies from early evening onwards, later coming to light.

1127(404). **G. myrtillana** (H. & W.) (*vacciniana* (Zell.))
O. On *Vaccinium myrtillus.*
L. 7-8. Between leaves spun flatly one above the other; probably overwinters unchanged in its cocoon.
P. 4-5. In a cocoon spun in leaf-litter or in the soil.
I. 5-6. Flies in evening sunshine.

1128(405). **Zeiraphera ratzeburgiana** (Ratz.)
O. On *Picea abies* or *P. smithiana*; less often on *Abies cephalonica, Pinus sylvestris* or *P. pinea.*
L. 5-6. In a young shoot; a shoot containing a larva retains the sheath for some time after the other shoots have shed it.
P. 6-7. In a flimsy cocoon amongst leaf-litter.
I. 7-8. Flies round the higher boughs at sunset and later comes to light.

1129(403). **Z. rufimitrana** (Herr.-Schäff.)
O. 7-4. Laid in small batches under the scales on a young shoot of *Abies alba, A. cephalonica* or *Pinus pinea.*
L. 4-6. In a spun terminal shoot; a shoot containing a larva retains the scale cap after other shoots have shed it. Later the larva moves to a fresh shoot which turns reddish brown.
P. 6-7. In a cocoon amongst leaf-litter.
I. 7. Flies at dusk round the tops of the trees and later comes to light.

1130(406). **Z. isertana** (Fabr.) (*corticana* sensu Hübn., 1811-13)
O. On *Quercus* spp.
L. 5-6. In spun leaves or in a folded leaf; occasionally in a spongy oak gall.
P. 6-7. In the larval habitation or spun up in a crevice in bark.
I. 7-8. Rests on the trunks but is readily disturbed; flies at dusk and later comes to light.

1131(402). **Z. diniana** (Guen.)
O. 7-5. Laid in small batches on the twigs of *Larix* and *Pinus* spp.; occasionally on *Abies alba.*

L. 5-6. In needles spun into a neat tube which it eats evenly from the top downwards; each larva constructs four to five such tubes.
P. 6-7. In a tough cocoon spun amongst fallen needles or other leaf-litter.
I. 7. Flies high round the trees in late afternoon and later comes to light.

1132(384). **Gypsonoma aceriana** (Dup.)
O. On *Populus* spp.
L. 5-6. In a shoot or leaf-stalk; an erect conical tube of frass at the entrance hole indicates the feeding-place.
P. 6. In a cocoon in a crevice of bark or amongst leaf-litter.
I. 7. Rests on the trunks; flies towards sunset and later occasionally comes to light.

1133(387). **G. sociana** (Haw.) (*neglectana* (Dup.))
O. On *Populus nigra* or *P. tremula*; less often on *Salix* spp.
L. 9-5. Before hibernation mining a twig; in the spring in the expanding leaf-buds or, when on *Salix*, in the catkins.
P. 6. In a cocoon in a crevice of bark or amongst leaf-litter.
I. 7. Rests on the trunks during the daytime; flies in the evening and later comes to light.

1134(386). **G. dealbana** (Frӧl.)
O. On *Crataegus, Quercus* spp., *Populus* spp., *Corylus* or *Salix* spp.
L. 9-6. After hibernation in a shoot or spun leaves; when on *Populus* or *Salix*, in the catkins.
P. 6-7. In a cocoon in the larval habitation or the soil.
I. 7-early 8. Flies at dusk and later comes to light.

1135(388). **G. oppressana** (Treits.)
O. On *Populus nigra* or *P. alba*.
L. 9-early 5. Prior to hibernation, beneath a silk web on the underside of a leaf; in the spring, in the buds, making a tube of frass and silk which at first projects over the boring but later lies flat along the surface of the twig.
P. 5-6. In a flimsy cocoon in the soil or, occasionally, a crevice of bark.
I. 6-7. Rests on the trunks; flies towards sunset and later comes to light.

1136(385). **G. minutana** (Hübn.)
O. On *Populus* spp.
L. 5-6. Between two leaves spun one above the other.
P. 6. In the feeding place.
I. 7. Flies rather high round the poplars in afternoon sunshine; later occasionally comes to light.

1137(413). **G. nitidulana** (L. & Z.)
O. 6. Laid singly on the underside of a leaf of *Populus tremula*; hatches in 16 days.
L. 7-4. Between spun leaves or in the folded edge of a leaf; hibernates in a similar spinning which falls to the ground.

P. 4-5. In a cocoon spun amongst dead leaves in April.
I. 6. Flies in the late afternoon and early evening.

1138(399). **Gibberifera simplana** (Fisch.v.Rösl.)
O. On *Populus tremula*, usually a young tree.
L. 8-9. In a spun leaf of a terminal shoot.
P. 9-5. In the turned-down edge of a leaf which falls to the ground in the autumn.
I. 6. Flies from early evening till dusk.

1139(439). **Epiblema cynosbatella** (Linn.) (*tripunctana* (Fabr.))
O. On *Rosa* or *Rubus* spp.
L. 4-5. In flower buds or a young shoot, or between spun leaves.
P. 5. In the feeding place.
I. 5-6. Flies from late afternoon onwards and occasionally comes to light.

1140(390). **E. uddmanniana** (Linn.)
O. 6-7. Laid singly on the upperside of a leaf of *Rubus* spp.
L. 7-5. After hibernation, in a terminal shoot spun into a large bunch.
P. 5-6. In the feeding place or in a spun leaf or moss.
I. End 6-7. Flies from dusk onwards, readily coming to light.

1141(392). **E. trimaculana** (Haw.)
O. On *Crataegus*.
L. 4-5. In spun leaves, preferring terminal shoots high in the tree.
P. 5-6. In the feeding place; may be obtained by beating.
I. 6. Flies at dusk and later comes to light.

1142(391). **E. rosaecolana** (Doubl.)
O. On *Rosa*, especially *R. rubiginosa*; often on a cultivar.
L. 5-6. In a spun young shoot, boring into the heart.
P. 6. In the feeding place or in a flimsy cocoon spun in leaf-litter.
I. 6-7. Flies from dusk onwards and comes to light.

1143(393). **E. roborana** ([D. & S.])
O. On *Rosa*, including cultivars.
L. 5-6. In a tightly spun shoot, sometimes attacking flower buds.
P. 6. In the feeding place.
I. End 6-early 8. Flies from evening onwards and comes readily to light.

1144(394). **E. incarnatana** (Hübn.)
O. On *Rosa*, especially *R. pimpinellifolia*.
L. 5-6. In spun leaves.
P. 6-7. In the feeding place.
I. 6-early 9. Flies from late afternoon onwards and comes readily to light.

1145(395). **E. tetragonana** (Steph.)
O. On *Rosa*.
L. 5-6. In spun leaves.
P. 6-7. In a flimsy cocoon amongst leaf-litter.
I. 7. Flies swiftly from 1700 hrs onwards in woodland rides and occasionally comes to light.

1146(431). E. **grandaevana** (L. & Z.)
O. On *Tussilago* or *Petasites* spp.
L. 8-5. Inside a root or in a silken tube on its surface.
P. 6. In the larval habitation.
I. 7. Flies at dusk.

1147(432). E. **turbidana** (Treits.)
O. On *Petasites hybridus*.
L. 8-4. In the roots or in a silk-lined tunnel in the crown; can be obtained in March by digging up the crowns of butterbur.
P. 5-6. In a cocoon just below the surface of the soil close to the roots of the foodplant.
I. End 6-7. Flies in late afternoon and again at sunset on warm evenings.

1148(430). E. **foenella** (Linn.)
O. On *Artemisia vulgaris* or *A. abrotanum*.
L. 8-5. In the roots, overwintering full-fed in the feeding place; prior to hibernation the larva eats partially through the stems just above the rootstock, covering the hole in the centre of the stem with silk interwoven with fragments of the stem. The stems thus eaten break off readily and indicate the presence of the larva (cf no. 1340).
P. 6. In the hibernaculum. The pupa is very active and can travel rapidly along the gallery by wriggling.
I. 7.

1149(434). E. **scutulana** ([D. & S.]) (*pflugiana* sensu Haw.)
O. On *Carduus nutans* or *Cirsium vulgare*.
L. 8-4. In the stem and roots.
P. 4-5. In the upper part of the larval mine.
I. 5-6. Rests by day on the foliage of its foodplants; flies in the evening and later comes to light.

1150(435). E. **cirsiana** (Zell.)
O. On *Cirsium palustre* or *Centaurea nigra*.
L. 8-4. In the stem and roots.
P. 4-5. In the upper part of the larval mine.
I. 5-6. Habits similar to those of *E. scutulana* and possibly only a form of that species.

1151(436). E. **cnicicolana** (Zell.)
O. On *Pulicaria dysenterica*.
L. 7-4. In the lower part of the stem.
P. 4-5. In a cocoon amongst the roots of the foodplant.
I. Mid 5-6. Flies over the foodplant in the evening.

1152(433). E. **farfarae** (Fletch.)
O. On *Tussilago;* occasionally on *Petasites fragrns.*
L. 7-4. At first in the roots, moving up into the flower-stem in the spring.
P. 4-5. In a cocoon amongst the roots of the foodplant.
I. 5-6. Flies in afternoon sunshine onwards till dusk, and also rests on leaves in full sunshine.

1153(437). **E. costipunctana** (Haw.) (*trigeminana* (Steph.))
O. On *Senecio jacobaea*.
L. 6-7;9-4. Summer brood in the stem; autumn brood in the rootstock, leaving in December and overwintering in a cocoon spun amongst leaf-litter.
P. 4;7. In the cocoon; summer brood sometimes in the stem.
I. 5-7;7-8. Rests on the foodplant by day; flies in late afternoon sunshine and later occasionally comes to light.

1154(429). **Pelochrista caecimaculana** (Hübn.)
O. On *Centaurea nigra*; probably other *Centaurea* spp.
L. 9-5. In the roots.
P. 6. In the larval habitation.
I. 7. Readily disturbed by day.

1155(401). **Eriopsela quadrana** (Hübn.)
O. 5-early 6. Laid singly on a leaf of *Solidago virgaurea*; hatches in about 14 days.
L. 6-4. In folded radical leaves, turning the edges upwards. Constructs a cocoon in late July amongst leaf-litter in which it overwinters unchanged.
P. 4-5. In the cocoon,
I. 5-early 6. Flies in late afternoon and later comes to light.

1156(422). **Eucosma aspidiscana** (Hübn.)
O. On *Solidago virgaurea*; occasionally on *Aster amellus* or *Crinitaria*.
L. 8-4. In the flower-head, later burrowing into the stem; spins a cocoon in the soil or under leaf-litter in late July, and overwinters therein.
P. 4. In the cocoon.
I. End 4-6. Easily disturbed in sunshine; flies at sunset and later comes to light.

1157(423). **E. heringiana** (Jäckh) (*rubescana* sensu auctt.)
O. On *Aster tripolium*.
L. 8-4. In spun flowers; overwinters from August in a cocoon spun in tidal litter at the base of the foodplant.
P. 5-6. In the cocoon.
I. 6-7. Flies actively for about half an hour at sunset.

1158(424). **E. conterminana** (Herr.-Schäff.)
O. 8. Laid singly or in small batches on a bract of a flower-head of *Lactuca virosa, L. serriola* or cultivated lettuce.
L. 8-5. At first, more or less exposed on the flower-head; later, burrowing into the seed-head. Overwinters from September in a cocoon spun in the soil or amongst leaf-litter.
P. 6. In the cocoon.
I. 7-8. Readily disturbed by day; flies towards sunset and later comes to light.

1159(421). **E. tripoliana** (Barr.) (*aemulana* sensu auctt.)
O. On *Aster tripolium*.
L. 9-5. In spun flowers and seed-heads, depositing frass amongst the pappus. Overwinters from late October in a cocoon spun in tidal litter or in the ground.

P. 6-7. In the cocoon.
I. 7-8. Flies in late afternoon and later comes to light.

1160(420). **E. aemulana** (Schläg.) (*latiorana* (Herr.-Schäff.))
O. On *Solidago virgaurea*.
L. 8-6. On the unripe seeds, moving from flower to flower without spinning them together. Overwinters from September in a cocoon spun in leaf-litter or in the soil.
P. 6. In the cocoon.
I. 7-8. Flies from late afternoon; later comes to light.

1161(418). **E. maritima** (H. & W.)
O. On *Artemisia maritima*.
L. 7-6. On the flowers and leaves, spinning them together into a tube. Overwinters from October in a cocoon spun in ground-litter or in the earth.
P. 6-7. In the cocoon.
I. 7-8. Flies from late afternoon onwards.

1162(–). **E. metzneriana** (Treits.) *Ent.Rec.* **89**:329(1977)
O. *On *Artemisia absinthium* or *A. vulgaris*.
L. *8-5. In the tip of a shoot which it aborts, causing a swelling.
P. *5-6. In the lower part of the stem.
I. 6-7. One specimen taken near Cambridge in 1977.

1163(416). **E. campoliliana** ([D. & S.]) (*nigromaculana* (Haw.))
O. On *Senecio jacobaea*.
L. 8-5. In the seeds and sometimes burrowing into the stem. Overwinters in a cocoon in the soil.
P. 5-6. In the cocoon.
I. 6-7. Flies from late evening and later comes to light.

1164(415). **E. pauperana** (Dup.)
O. On *Rosa canina*.
L. 6. On the unripe hips, spinning them together.
P.
I. 4-5. May be disturbed by beating the foodplant; flies in warm evenings about 1700 hrs and later comes to light.

1165(419). **E. pupillana** (Clerck)
O. On *Artemisia absinthium*.
L. 9-6. At first, under the bark of a stem; later, in the rootstock.
P. 7. In the larval burrow or amongst the roots of the foodplant.
I. 7-8. Rests in the foliage of the foodplant on warm afternoons and flies at dusk.

1166(427). **E. hohenwartiana** ([D. & S.]) (*scopoliana* Haw.)
O. On *Centaurea nigra, C. scabiosa* or *Serratula*.
L. 8-5. In the flower-head or seed-head. Overwinters from September in a cocoon spun in leaf-litter or the soil.
P. 5-early 7. In the cocoon.
I. 6-8. Flies towards dusk and later comes to light.

1167(426). **E. fulvana** (Steph.)
O. On *Centaurea scabiosa*.
L. 8-5. Life history similar to that of *E. hohenwartiana*.
P. 5-early 7.
I. 6-8. Possibly only a form of the preceding species.

1168(425). **E. cana** (Haw.)

O. On *Cirsium vulgare, Carduus* spp. or *Centaurea nigra*.

L. 8-5. In a flower-head, feeding on the pith and immature seeds. Overwinters from September in a cocoon spun amongst leaf-litter or in the soil.

P. 5-7. In the cocoon.

I. 6-8. Flies from late afternoon and later comes to light.

1169(428). **E. obumbratana** (L. & Z.) (*expallidana* Meyr. et auctt.)

O. On *Sonchus arvensis*.

L. 8-5. In a flower-head on the developing seeds. Overwinters from September in a cocoon in the soil.

P. 5-6. In the cocoon.

I. 7-8. Flies in the evening and later comes to light.

1170(–). **Foveifera hastana** (Hübn.)

A European species, specimens of which were reputed to have been taken in Devon in June, 1820.

1171(417). **Thiodia citrana** (Hübn.)

O. On *Achillea millefolium, Artemisia campestris* or *Anthemis cotula*.

L. 8-5. In the flower-head, often spinning two or three flowers together. Overwinters from September in a cocoon spun amongst leaf-litter.

P. 6. In the cocoon.

I. 7. Rests on flowers of the foodplants in afternoon sunshine; flies at dusk and later comes to light.

1172(353). **Spilonota ocellana** ([D. & S.])

O. 7-8. Laid singly or in small batches generally on the upperside of leaves of a wide range of foodplants, including many rosaceous trees and shrubs, *Salix, Hippophae, Myrica, Euphorbia paralias*, etc. Hatches in 7-10 days.

L. 8-6. Before hibernation, in an open-ended silk tube on the underside of a leaf; in spring first eating into the buds and later in a loose, untidy spinning.

P. 6. In the larval habitation.

I. 6-8. Flies at dusk and later comes to light.

1173(354). **S. laricana** (Hein.) (*lariciana* missp.)

O. 7-8. On *Larix*; sometimes on *Picea sitchensis*.

L. 8-6. Before hibernation, mining spun needles; in the spring in expanding buds.

P. 6. In the bud in which it last fed.

I. 6-8. May be a form of the preceding species.

1174(366). **Clavigesta sylvestrana** (Curt.)

O. On *Pinus pinaster* or *P. pinea*; possibly on *P. sylvestris*.

L. 8-6. In the shoots, buds and male flowers.

P. 5-6. In a tough white silken cocoon at the base of a flower near the larval habitation.

I. 6-7. Flies in the later afternoon and later occasionally comes to light.

1175(365). **C. purdeyi** (Durr.)

O. 8-9.	Laid singly near the tip of a needle of *Pinus sylvestris* or *P. nigra* var. *maritima*.
L. 9-7.	Mines needles in autumn and early spring; later on in the base of the new growth of needles, causing them to fall, the empty sheaths indicating the presence of a larva nearby.
P. End 6-7.	In a loosely spun cocoon attached to the needles.
I. End 7-9.	Flies in the evening and later comes to light; may be obtained in the daytime by beating.

1176(363). **Blastesthia posticana** (Zett.)

O.	On *Pinus sylvestris*, on lateral shoots of smaller trees.
L. 9-4.	In a shoot, aborting its growth; in spring in a bud. An infested bud remains dull brown, lacking the reddish tinge of developing buds.
P. 4-5.	In the excavated bud.
I. 5-6.	Flies towards evening.

1177(361). **B. turionella** (Linn.) (*turionana* (Haw.))

O.	On *Pinus sylvestris* or, occasionally, *P. contorta*.
L. 8-3.	In a bud, usually the central bud of the topmost shoot, aborting the growth.
P. 4-5.	In the excavated bud.
I. 5-6.	Flies from late afternoon onwards, and later comes to light.

1178(359). **Rhyacionia buoliana** ([D. & S.])

O. 7-8.	Laid singly or in small batches on the young growth, close to the terminal whorl, of *Pinus sylvestris, P. contorta* or *P. nigra*.
L. 9-6.	In a developing bud, causing resinous exudation.
P. 6-7.	In a silken cocoon in the feeding place.
I. End 6-mid 8.	Flies in the late evening and later comes to light.

1179(358). **R. pinicolana** (Doubl.)

O.	On *Pinus sylvestris*.
L. 4-5.	In a shoot, causing resinous exudation and distortion.
P. 6-7.	In a silken cocoon covered with frass in the larval habitation.
I. 7-8.	Flies at dusk and later comes to light; may be obtained in the daytime by beating young pines.

1180(360). **R. pinivorana** (L. & Z.)

O.	On *Pinus sylvestris*.
L. 8-4.	Overwinters in a hollowed-out bud; in spring in a shoot, preferring a lateral shoot.
P. 5-6.	Usually in the feeding-place; occasionally in spun needles.
I. End 5-early 7.	Flies in the evening and later comes to light; occasionally flies also in bright sunshine and is easily disturbed by day.

1181(364). **R. duplana logaea** Durr.

O.	Laid singly or in small batches on a needle of *Pinus sylvestris* or *P. contorta* var. *latifolia*.
L. 5-8.	In a bud or a young shoot, causing it to shrivel and break off.

P. 8-3. In the larval habitation or in a cocoon spun in an angle formed by a twig.
I. 4-early 5. Flies from mid afternoon onwards; easily disturbed in the daytime.

1182(362). **Petrova resinella** (Linn., 1758)
O. 6-7. On a young shoot of *Pinus sylvestris*, usually near a bud-whorl; hatches in about 21 days.
L. 7-4, living two years. In a large resinous gall on a twig; in the second autumn it spins a cocoon in the gall and overwinters therein.
P. 4-5. In the cocoon in the gall.
I. 6. Common in odd-numbered years.

1183(496). **Cryptophlebia leucotreta** (Meyr.)
O.
L. In oranges or other citrus fruits; a sunken brown patch with a hole in the centre may indicate the presence of a larva.
P. In a tough cocoon spun in debris or the soil.
I. An African species sometimes imported in fruit.

1184(546). **Enarmonia formosana** (Scop.) (*woeberiana* ([D. & S.]))
O. 7-8. Laid singly or in small batches on a smooth area of the bark of various rosaceous trees, especially mature *Malus* and ornamental cherries.
L. 9-5. In the bark, exuding frass.
P. 5-6. In the feeding place.
I. 6-7. Emerges at about 1000 hrs and rests on the trunk till noon; flies in afternoon and evening sunshine and later occasionally comes to light.

1185(396). **Eucosmomorpha albersana** (Hübn.)
O. On *Lonicera periclymenum*.
L. 8-4. In a leaf folded into a pod or between spun leaves; overwinters from September in a cocoon spun amongst leaf-litter.
P. 4-5. In the cocoon.
I. 5-6. Flies actively or rests on the foliage of the foodplant in afternoon sunshine; later occasionally comes to light.

1186(558). **Selania leplastriana** (Curt.)
O. Laid singly on the stem of *Brassica oleracea*.
L. 9-5. In the skin and side shoots, exuding frass.
P. 6. In the feeding place or in a cocoon spun amongst surface litter.
I. 7. Apparently confined to coastal cliffs where it is easily disturbed from the foodplant and flies in afternoon sunshine.

1187(460). **Lathronympha strigana** (Fabr.) (*hypericana* (Hübn.))
O. On *Hypericum* spp., especially *H. perforatum*.
L. 4-5;7. In a spun terminal shoot. in favourable seasons only, a second generation occurs, feeding on the flowers and seeds.
P. 5-6.;7-8. In the soil or under moss.
I. End 5-6;8. Flies at sunrise and in the evening; occasionally comes to light.

1188(553). **Collicularia microgrammana** (Guen.)

O.	On *Ononis repens*.
L. 8-9.	In the green seed-pods.
P. 5-6.	In a cocoon of silk and earth attached to recumbent stems of the foodplant or in detritus.
I. End 5-7.	Flies at dusk and later comes to light.

1189(528). **Strophedra weirana** (Dougl.)

O.	On *Fagus*.
L. 8-9.	Between two leaves spun one above the other, making pale blotches.
P. 10-5.	Between the two leaves, which eventually fall to the ground.
I. 6.	Flies at sunrise and in sunshine in the afternoon and evening.

1190(529). **S. nitidana** (Fabr.)

O.	On *Quercus* spp.
L. 7-9.	Between two leaves spun one above the other, making pale blotches.
P. 10-4.	Between the two leaves, which eventually fall to the ground.
I. 5-6.	Flies in sunshine rather high up round oak-trees from 1500-1830 hrs.

1191(532). **Pammene splendidulana** (Guen.)

O.	On *Quercus* spp.
L. 6-7.	In a silken gallery between two leaves spun flatly together, causing window-like patches, never eating the outer epidermis.
P. 8-3.	In a silken cocoon under bark, in rotten wood, in an old oak-apple, or a spongy gall, including those which have fallen. Some larvae do not pupate till early spring.
I. 4-early 6.	Flies high up round oak-trees in afternoon and evening sunshine.

1192(–). **P. luedersiana** (Sorh.) *Ent.Rec.* **88**:88(1976)

O.	
L. 7.	Probably on *Myrica*, in spun leaves.
P.	
I. 5-6.	

1193(438). **P. obscurana** (Steph.)

O.	On *Betula* spp.
L. 7.	In the catkins.
P.	
I. 5-6.	Flies in afternoon and evening sunshine.

1194(–). **P. agnotana** Reb. *Ent.Rec.* **75**:13-14(1963)

O.	
L.	*Under the bark of *Crataegus*.
P.	
I. 4.	One specimen taken in Surrey in 1961.

1195(533). **P. inquilana** Fletch. (*fimbriana* (Haw.) nec (Thunb.))

O.	
L. 6-8.	In a young oak-apple or spongy gall on *Quercus* spp., sometimes exuding frass.

P. 9-3. In a silk cocoon in the larval habitation, or in an old gall, dead twigs, under bark or between dead leaves.
I. End 3-5. Flies in afternoon sunshine and later occasionally comes to sallow bloom and light.

1196(534). **P. argyrana** (Hübn.)
O.
L. 6-3. In a gall on *Quercus* spp., preferring the spongy galls. Overwinters from September in its cocoon.
P. 3. In a cocoon under bark or moss, in an old gall or sometimes in the gall in which the larva fed.
I. 4-5. Rests by day on the trunks of oak; flies from late afternoon onwards and later comes to light.

1197(535). **P. albuginana** (Guen.) (*gallicolana* sensu auctt.)
O.
L. 9-3. In a spongy gall on *Quercus* spp. Overwinters from October in its cocoon.
P. 3-4. In a cocoon in the larval habitation, an old gall or a crevice in bark.
I. 6. Rests by day on the foliage, from which it may be dislodged by beating; flies high round oaks in late afternoon and evening and later comes to light.

1198(–). **P. suspectana** L. & Z. *Ent.Gaz.* **27**:148(1976)
O. *On *Fraxinus.*
L. *Under the bark.
P.
I. One specimen was captured in a "sticky trap" near Cambridge in about 1975 and a second in 1978.

1199(539). **P. spiniana** (Dup.)
O. On *Prunus spinosa* or *Crataegus.*
L. 5-6. In spun flowers.
P. 7. In the larval spinning.
I. 8-9. Flies in sunshine over the tops of its foodplants; best obtained by beating on a dull day.

1200(540). **P. populana** (Fabr.)
O. On *Salix* spp., especially *S. caprea, S. repens* and *S. viminalis.*
L. 5-6. On the shoots, from a silk-lined spinning incorporating withered fragments of leaf.
P. 6-7. In a cocoon spun under loose bark or in a dead stem or rotten wood.
I. 7. Rests on the foliage of its foodplants in the afternoon.

1201(542A). **P. aurantiana** (Staud.) *Ent.Gaz.* **9**:60(1955); *Ent.Rec.* **71**:222, 246(1959); **72**:34-35, and 247(1960)
O. On *Acer pseudoplatanus.*
L. 8-9. In the seeds.
P. In captivity, in dead wood or bark.
I. 7-8. Flies high up in the afternoon; later comes to light.

1202(541). **P. regiana** (Zell.)
O. On *Acer pseudoplatanus* or *A. platanoides*; probably also on *A. campestre.*

L. 8-4. In the seeds; overwinters from September in its cocoon.
P. 4-5. In a cocoon spun under flakes of bark near the base of the host tree.
I. 5-7. Emerges between 0800 and 1000 and later ascends high up on the trees, though some may remain low down. Occasionally taken at light.

1203(542). **P. trauniana** ([D. & S.])
O. On *Acer campestre*.
L. 8-4. In the seeds; overwinters from September in its cocoon.
P. 4. In a cocoon under bark at the base of the host tree.
I. 5-6. Flies in sunshine round the tops of the maples or rests amongst the foliage.

1204(536). **P. fasciana** (Linn.) (*juliana* Steph.))
O. Laid singly or in small batches beside a vein on a leaf of *Quercus* spp. or *Castanea*, close to the fruits.
L. 8-5. In an acorn or chestnut, which falls and is vacated before those containing larvae of *Cydia splendana*. Overwinters in its cocoon from October.
P. 5-6. In a cocoon spun in rotten wood or surface litter, under moss or bark, or occasionally in old spongy oak galls.
I. 6-7. Males fly at sunrise and in afternoon sunshine; females sometimes come to light.

1205(537). **P. herrichiana** (Hein.) (*nimbana* (Herr.-Schäff.))
O. On *Fagus*.
L. 8-4. In the mast. Overwinters from October in its cocoon.
P. 4-5. In a cocoon in situations similar to those selected by the last species.
I. 5. Considered by some to be a form of *P. fasciana*.

1206(530). **P. germmana** (Hübn.) (*germarana* (Hübn.))
O.
L. Life history not properly known; variously stated to feed in the fruits of *Prunus domestica* or in the shoots of *Quercus* spp. or *Crataegus*.
P.
I. 6. Flies early in the morning and again in the afternoon and evening about the tops of oak-trees.

1207(531). **P. ochsenheimeriana** (L. & Z.) *Ent.Rec.* **90**:67-68(1978)
O. On *Abies grandis*; possibly also on *Pinus sylvestris* or *Picea abies*.
L. 7-8. Possibly in buds attacked by larvae of *Epinotia nigricana*, or in dead or dying twigs; in aborted buds of *A. grandis*.
P. 4-5. In a cocoon in the larval habitation.
I. 5-6. Flies round the tips of the branches of the foodplants in late afternoon.

1208(538). **P. rhediella** (Clerck)
O. Laid singly on the underside of a leaf close to a fruit of *Crataegus* or, less often, *Swida, Malus, Pyrus, Prunus domestica* or *P. cerasus*.
L. 6-3. In spun flowers or on the surface of a fruit; overwinters from July in its cocoon.

P. 4. In a cocoon in bark, dead wood or the soil.
I. 5-6. Flies in sunshine round the tops of hawthorns.

1209(555). **Cydia caecana** (Schläg.)
O. On *Onobrychis*.
L. 8-5. In the stem.
P. 6. In the mined stem.
I. 7. Flies round the foodplant in afternoon and evening sunshine.

1210(563). **C. compositella** (Fabr.)
O. Laid singly on leaves of *Trifolium* spp., especially *T. pratense* and *T. repens*.
L. 6-7;8-4. First generation in the stem; second in spun terminal leaves or a flower-head, overwintering full-fed.
P. 4-6;7-8. In a cocoon spun amongst leaf-litter.
I. 4-5;8. Males fly in afternoon sunshine, females towards sunset.

1211(561). **C. internana** (Guen.)
O. On *Ulex europaeus*.
L. 7-3. In the pods, overwintering from September in its cocoon.
P. 3-4. In a cocoon in the pod in which the larva fed or amongst the leaves of the foodplant.
I. 4-6. The male flies actively in sunshine; the female flies less readily.

1212(562). **C. pallifrontana** (L. & Z.)
O. On *Astragalus glycyphyllos*.
L. 7-4. In the pods, overwintering from August in its cocoon.
P. 4-5. In a cocoon in the pod or the soil; in captivity in rotten wood.
I. 6-7. Males especially fly over the foodplant in afternoon sunshine.

1213(554). **C. gemmiferana** (Treits.)
O. On *Lathyrus sylvestris*.
L. 7-4. In a spun leaf or between two leaves spun together, overwintering from the end of August in its cocoon.
P. 4. In a cocoon spun amongst leaf-litter.
I. End 5-6. Flies in sunshine.

1214(547). **C. janthinana** (Dup.) (*ianthinana* missp.)
O. Laid singly on a berry of *Crataegus*.
L. 9-5. In the berries, spinning two or three together; overwinters from October in its cocoon.
P. 5-6. In a cocoon under loose bark, in rotten wood, or amongst leaf-litter.
I. 7. Flies in afternoon sunshine up to dusk in warm weather.

1215(543). **C. tenebrosana** (Dup.) (*roseticolana* Zell.)
O. On *Rosa* spp., especially *R. canina*.
L. 8-4. In a hip, at first causing brownish stripes on the surface, later causing the berry to wrinkle and turn purple; overwinters from September in its cocoon.
P. 4-6. In a cocoon in dead wood or under bark.
I. 6-7. Flies high over the roses in sunshine and later comes to light.

1216(545). **C. funebrana** (Treits.)

O. 6-7. Laid singly at the base of a fruit-stalk of *Prunus* spp., especially *P. spinosa* or *P. domestica*; hatches in 10 days.

L. 7-4. In the fruit, leaving a conspicuous entrance-hole surrounded by frass, generally near the stalk; overwinters full-fed from late August or early September in a cocoon spun in a crevice of bark, under moss, in the soil or, in captivity, in rotten wood.

P. 4-5. In the cocoon.

I. 6-7. Rests by day high in the trees; flies at night and comes to light.

1217(544). **C. molesta** (Busck)

O. Laid singly on a leaf or twig of *Prunus* spp., especially *P. persica* or *P. armeniaca, Pyrus, Malus* or other Rosaceae; hatches in 7-30 days, depending on temperature.

L. 4-6;7-3. In a twig or fruit or on the leaves in two or more generations;autumn larvae overwinter.

P. 3-4;7. In a whitish silken cocoon on a twig or fruit, under bark or in the earth.

I. 3-9. A pest species often imported with fruit and unlikely to become established in Britain.

1218(573B). **C. prunivorana** (Rag.) (*lobarzewskii* sensu auctt.) *Ent.Rec.* **70**: 70-71(1958); **88**:268(1976)

O. Laid singly on a fruit of *Prunus* spp. or *Malus*.

L. 7-9. In the fruit.

P. 4. In a cocoon under bark or in rotten wood.

I. 5-6. Possibly bivoltine.

1219(568). **C. lathyrana** (Hübn.) (*scopariana* (Herr.-Schäff.))

O. On *Genista tinctoria*.

L. 6-3. On unexpanded flower-buds, spun flowers or young shoots; overwinters full-fed from July in a cocoon spun at the base of the foodplant.

P. 3-4. In the cocoon.

I. 4-5. Flies in sunshine for about two hours at midday.

1220(560). **C. jungiella** (Clerck) (*perlepidana* (Haw.))

O. On *Lathyrus montanus* or *Vicia sepium*.

L. 6-3. In a pod or between tightly spun leaves; full-fed in August.

P. 3-4. In a cocoon in the soil or amongst leaf-litter.

I. 4-5. Flies in afternoon sunshine and at night, when it comes to light. There is sometimes a small second generation in the south during August.

1221(565). **C. lunulana** ([D. & S.]) (*dorsana* sensu auctt.)

O. On *Lathyrus montanus*; probably also *L. pratensis, Vicia cracca* or *Pisum sativum*.

L. 6-4. When young, mining the stem or in spun leaves; later in a pod feeding on the seeds, causing discoloration near the base and arresting development; overwinters full-grown from August in a cocoon in the pod or in the soil or amongst litter.

P. 4-5. In the cocoon.

I. 5-6. Flies in sunshine, especially during the afternoon.

1222(564). **C. orobana** (Treits.)

O. On *Vicia* spp., especially *V. sylvatica* or *V. cracca*, or on *Lathyrus pratensis* or *L. palustris*.

L. 8-4. In the pods on the seeds; overwinters full-fed from September in a cocoon spun in the soil or amongst leaf-litter.

P. 4-5. In the cocoon.

I. 7. Flies in late afternoon sunshine and later comes to light.

1223(570). **C. strobilella** (Linn.)

O. On *Picea abies*.

L. 8-4. In a cone on the seeds, sometimes causing slight resinous exudation.

P. 4-5. In the cone; fallen cones collected in winter generally produce adults.

I. 5. Flies in afternoon sunshine round the tops of the spruces.

1224(556). **Cydia succedana** ([D. & S.]) (*ulicetana* (Haw.))

O. On *Ulex, Genista, Sarothamnus* or *Lotus*.

L. 6;9-4. On the unripe seeds in the pods, moving from pod to pod. The second generation overwinters full-fed from October in a cocoon spun in leaf-litter or the soil.

P. 4;6-7. In the cocoon.

I. 5;6-7. Males fly actively in sunshine, females towards sunset. Univoltine in Scotland, flying from May to July.

1225(557). **C. servillana** (Dup.)

O. 5-6. Laid singly on a bud of *Salix caprea* or *S. cinerea*.

L. 8-4. In a slender gall on a one-year-old twig; overwinters full-fed from October in a cocoon in the gall; the exit-hole, situated just above the bud and covered by silk mixed with reddish frass, is generally conspicuous. Galls collected in winter will produce adults in about six weeks if kept in a warm room.

P. 4-5. In the cocoon in the gall.

I. 5-6. The male flies in afternoon sunshine and both sexes in the evening.

1226(566). **C. nigricana** (Fabr.)

O. Laid singly on a leaf, stipule, sepal or developing pod of various wild or cultivated species of *Vicia* or *Lathyrus*, or on *Pisum sativum*.

L. 7-4. On the seeds in a pod; overwinters full-fed from September in a hibernaculum in the soil.

P. 5-6. In a cocoon spun closer to the surface of the soil than the hibernaculum.

I. 5-8. Has a long emergence-period, but a small second brood is possible. Flies in afternoon sunshine and later occasionally comes to light. A pest on *Pisum*.

1227(573A). **C. milleniana** Adamczewski (*deciduana* (Steuer); *zebeana* sensu auctt.)

O. 5-6. Laid singly on a twig of *Larix*.

L. 7-4. Living two years, in the second year making a resinous gall at a junction of twigs; tenanted galls are soft and spongy, old galls hard and woody.

P. 4-5. In the gall.
I. 5-6. Flies rather high round the larches in afternoon sunshine.

1228(552). **C. fagiglandana** (Zell.) (*grossans* (Haw.) nec (Thunb.))
O. On *Fagus*.
L. 8-4. On the mast, hollowing out the nut and packing it with frass; after the mast has fallen, it vacates the nut and constructs a cocoon in which it overwinters. The timing is variable, since larvae may sometimes be found in March crawling on the beech trunks in search of a site for pupation.
P. 5. In a cocoon spun in the soil, in litter, in rotten wood or under bark or moss.
I. 6-7. Freshly emerged moths rest on the trunks; they fly rather high in the evening and later come to light.

1229(557). **C. splendana** (Hübn.)
O. On *Quercus* spp., *Castanea* or occasionally, *Juglans*.
L. 8-5. In an acorn, chestnut or walnut; when full-fed in October, after the nuts have fallen, it spins a cocoon in the soil, leaf-litter, moss or bark, in which it overwinters.
P. 6. In the cocoon.
I. 7-8. Flies at dusk and later comes to light.

1230(550). **C. pomonella** (Linn.)
O. Laid, usually singly, on or near the calyx of a fruitlet or on the side of a young fruit of *Malus, Cydonia, Pyrus, Juglans, Castanea, Ficus* or *Sorbus aria*, hatching in 10-15 days.
L. 8-4. In the core, eating the flesh and seeds, or in spun berries, according to the foodplant. When full-fed in September or October, it overwinters in a cocoon spun under bark or in leaf-litter.
P. 4-5. In the cocoon.
I. 7-8, but some individuals, which emerge earlier, may give rise to a second generation of adults emerging 8-10. Rests on the trunks and foliage by day, flies at dusk and later comes to light.

1231(–). **C. amplana** (Hübn.)
Stated to be British by Curtis (1831), but no authentic British specimen is known.

1232(–). **C. inquinatana** (Hübn.)
Stated to be British by Stephens (1829), but no authentic British specimen is known.

1233(567). **C. leguminana** (L. & Z.)
O. 6-7. Laid on the bark of *Ulmus procera* near an excrescence; possibly also other trees such as **Acer pseudoplatanus* and *Fagus*.
L. 8-3. In decaying bark, especially on pollarded trees which have excrescences with soft bark.
P. 4-5. In the larval habitation or under loose bark.
I. 5-6. Flies freely in afternoon sunshine.

1234(571). **C. cognatana** (Barr.) (*pactolana* sensu Ford)

O. On *Pinus sylvestris*.

L. 8-4. In the bark, generally low down; reddish brown frass in the crevices indicates the presence of a larva.

P. 4-5. In the larval habitation.

I. 6. Flies round the tops of pines in afternoon sunshine.

1235(–). **C. pactolana** (Zell.) *Ent.Rec.* **78**:134(1966); **99**:76(1977)

O. *6. Laid in batches of two or three in a fissure of the bark of *Picea abies* or *Larix*, hatching in about 10 days; generally on trees 5-15 years old.

L. *7-4. In a gallery in the bark of a twig, exuding a small heap of yellowish frass at the entrance to the tunnel.

P. *4-5. In a silken cocoon amongst the frass near the entrance to the tunnel.

I. 6. Known in Britain only from two specimens taken in Hampshire in 1965 and 1976; comes to light.

1236(569). **C. cosmophorana** (Treits.)

O. Laid in a crevice in the bark of *Pinus sylvestris*.

L. 9-4. In a resinous nodule on the bark or in the one-year-old gall of *Petrova resinella* or an old feeding-place of *Dioryctria abietella*.

P. 4-5. In the larval habitation.

I. 5-6. Flies in sunshine from noon till late afternoon round the tops of pines.

1237(572). **C. coniferana** (Ratz.)

O. On *Pinus sylvestris* or *P. nigra* var. *maritima*.

L. 9-4. In a silk-lined tunnel under the bark, exuding reddish brown frass; often only one in a group of trees is infested.

P. 5-6. In the larval habitation.

I. 6-8. Apparently bivoltine in the south, univoltine in the north. Flies round pines in afternoon sunshine and again at dusk.

1238(573). **C. conicolana** (Heylaerts)

O. 6. Laid singly on one-year-old cones of *Pinus sylvestris* or *P. nigra* var. *maritima*, though several may be found on a single cone; cones growing in sunshine on the south side of a tree are preferred.

L. 7-3. In the cone on the seeds; when full-fed, it constructs a pupal chamber in the apical part of a scale and overwinters therein. There is no external evidence of the feeding.

P. 3-5. In the pupal chamber.

I. End 5-6. Flies in late morning and the afternoon round the tops of pines.

1239(559). **C. corollana** (Hübn.)

O. On *Populus tremula*.

L. *8-3. In old or occupied galls of the cerambycid beetle *Saperda populea* (Linn.) in twigs of the host plant.

P. *4-5. In the larval habitation.

I. 6-7. Taken *c*. 1850 in Huntingdonshire.

1240(548). **C. gallicana** (Guen.) (*rufillana* (Doubl.))

O.
L. 9-3. On loosely spun seeds of *Daucus, Angelica sylvestris, Heracleum sphondylium* and other umbellifers. Overwinters full-fed in a cocoon in the soil.
P. 4-6. In the cocoon.
I. 7-8. Flies in afternoon sunshine and again at dusk; rests on flowerheads of the foodplant.

1241(549). **C. aurana** (Fabr.)

O. On *Heracleum sphondylium*.
L. 8-4. On the seeds, spinning two or more together and living within. Overwinters full-fed from October in a cocoon spun in the soil.
P. 5-6. In the cocoon.
I. 6-7. Rests on flowerheads of the foodplant and flies in sunshine.

1242(513). **Dichrorampha petiverella** (Linn.)

O. On *Achillea* or *Tanacetum vulgare*.
L. 9-4. In the roots and rootstock.
P. 5. In an earthen cocoon amongst the roots of the foodplant.
I. 6-8. Rests on flowers of the foodplants; flies in sunshine in the afternoon and evening.

1243(514). **D. alpinana** (Treits.)

O. On *Leucanthemum vulgare*, including the cultivated varieties.
L. 9-6. In the rootstock.
P. 6-7. In an earthen cocoon attached to the rootstock.
I. 6-8. Rests on flowers of the foodplants; flies in sunshine and also at dusk.

1244(515). **D. flavidorsana** Knaggs

O. On *Tanacetum vulgare*.
L. 9-6. In the rootstock.
P. 6-7. In a cocoon in the larval habitation or attached to the rootstock.
I. 7-8. Rests on the foodplant; flies at dusk and occasionally earlier in sunshine.

1245(519). **D. plumbagana** (Treits.)

O. On *Achillea millefolium*.
L. 9-4. In the lower part of a flower-stalk, tunnelling down into the rootstock.
P. 5-6. In the larval habitation.
I. 6. Flies in afternoon sunshine and rests on the flower-heads of the foodplant.

1246(520). **D. senectana** (Guen.)

O. On *Leucanthemum vulgare* and *Tanacetum vulgare*.
L.
P.
I. 6. Frequents coastal areas where the vegetation is sparse; flies in the afternoon and at dusk.

1247(518). **D. sequana** (Hübn.)

O. On *Achillea millefolium* and *Tanacetum vulgare*.

L. 8-4. In the roots and rootstock.
P. 5-6. In the feeding place or in an earthen cocoon.
I. 6. Rests on flowers of the foodplant and flies towards sunset.

1248(523). **D. acuminatana** (L. & Z.)
O. On *Leucanthemum vulgare* or *Tanacetum vulgare*.
L. 10-4;7. In the rootstock, burrowing under the epidermis and eating the pith.
P. 4-5;7-8. In the larval habitation.
I. 5-6;8-9. The male flies freely in the late afternoon.

1249(524). **D. consortana** Steph.
O. On *Leucanthemum vulgare*.
L. 5-6. In a stem, causing a slight swelling and more or less aborting the growth; laternal shoots tend to develop which grow higher than the tenanted stem.
P. 6-7. In the larval feeding place or in an earthen cocoon.
I. 7-8. Flies round the foodplant in the afternoon.

1250(521). **D. simpliciana** (Haw.)
O. On *Artemisia vulgaris*.
L. 10-5. In the lower part of the stem and the rootstock.
P. 6. In the larval feeding place.
I. 7-9. Flies from sunset till dusk and later occasionally comes to light.

1251(517). **D. sylvicolana** Hein. (*alpestrana* sensu auctt.)
O. On *Achillea ptarmica*.
L. 8-4. Under the epidermis of the rootstock and roots, sometimes gregariously.
P. 5. In a cocoon amongst the roots of the foodplant.
I. 6-7. Flies only seldom by day; rests on the foodplant and scuttles down into the herbage if alarmed; may be obtained by sweeping or shaking the foodplant over a net.

1252(522). **D. montanana** (Dup.) (*tanaceti* P. & M.)
O. On *Achillea millefolium* or *Tanacetum vulgare*.
L. 10-4. In the rootstock.
P. 4-6. In an earthen cocoon.
I. 7-8. Flies round the foodplants in the evening; may be beaten out by day.

1253(516). **D. gueneeana** Obraztsov (*politana* sensu Ford)
O. On *Achillea millefolium* or *Tanacetum vulgare*.
L. 9-5. In the rootstock, sometimes gregariously.
P. 5-6. In the larval habitation.
I. 6-7. Rests on the foodplants; flies in afternoon sunshine and towards sunset.

1254(525). **D. plumbana** (Scop.)
O. On *Leucanthemum vulgare* or *Achillea millefolium*.
L. 8-4. In the rootstock, tunnelling down into the roots; finally it feeds on the surface of the roots under a web.
P. 4-5. In an earthen cocoon.
I. 5-6. Flies at all hours of day, especially in close, cloudy weather; also flies at dusk.

1255(527). **D. sedatana** Busck (*saturnana* sensu auctt.)
O. On *Tanacetum vulgare*.
L. 10-4. In the rootstock.
P. 5-6. In the larval habitation.
I. 6-7. The male flies in sunshine; both sexes fly at dusk.

1256(526). **D. aeratana** (P. & M.)
O. On *Leucanthemum vulgare*.
L. 10-4. In the roots.
P. 4-5. In an earthen cocoon.
I. 5-6. The male flies in afternoon sunshine and both sexes fly at dusk.

ALUCITIDAE

1257(824). **Alucita hexadactyla** (Linn.)
O. On *Lonicera periclymenum*.
L. 6-7. In flower-buds and mining the leaves.
P. 7-8. In a tough cocoon on or under the surface of the ground.
I. 8-4. Flies at dusk and after dark. Hibernates in outhouses or dense cover.

PYRALIDAE

1258(95). **Euchromius ocellea** (Haw.)
O.
L. On stored food and dead vegetation.
P.
I. 9-3. A scarce immigrant, perhaps only an accidental importation.

1259(96). **Chilo phragmitella** (Hübn.)
O. On *Phragmites* or *Glyceria maxima*.
L. 9-5. In a stem, frequently below water-level.
P. 6. In the stem near water-level.
I. 6-7. Flies at night and comes readily to light.

1260(97). **Acigona cicatricella** (Hübn.) *Ent.Rec.* **64**:160(1952)
O. *On *Scirpus lacustris*.
L. *5-6. In a stem.
P. *6-7.
I. 7. A rare immigrant to the south-east; one specimen was taken many years ago, another at light in 1951.

1261(92). **Calamotropha paludella** (Hübn.)
O. On *Typha latifolia* or, occasionally, *T. angustifolia*, on the edge of a leaf.
L. 9-6. At first mining the lower part of a leaf; after winter, feeding in a dead stem, sometimes gregariously.
P. 6-7. In a dead stem.
I. 7-8. Comes to light.

1262(69). **Chrysoteuchia culmella** (Linn.) (*hortuella* (Hübn.))
O.
L. 9-4. On stem-bases of grasses, without making a silken gallery.
P. 5. In a cocoon on or under the surface of the ground, in which the larva has overwintered.
I. 6-7. Comes readily to light.

1263(63). **Crambus pascuella** (Linn.)
O.
L. On grasses, especially on *Poa* spp.
P.
I. 6-7.

1264(–). **C. leucoschalis** (Hamps.) *Ent.* **92**: 176(1959)
A native of the drier parts of Africa; one was taken at Plymouth in 1920.

1265(65). **C. silvella** (Hübn.)
O.
L. On *Carex* spp.
P.
I. 6-8. Frequents boggy ground; flies at dusk and after dark.

1266(62). **C. uliginosellus** Zell.
O.
L. Probably on grasses.
P.
I. 6-7. Frequents boggy ground; flies at dusk and after dark.

1267(64). **C. ericella** (Hübn.)
O.
L. *On *Corynephorus* and *Aria* spp.
P.
I. 6-7. Frequents heaths and moors, generally between 600 and 750 metres; flies in the late afternoon.

1268(61). **C. hamella** (Thunb.)
O.
L.
P.
I. 7-8. Frequents sandy heaths; flies late at night.

1269(66). **C. pratella** (Linn.) (*dumetella* (Hübn.))
O. On grasses, especially *Deschampsia* spp.
L. 9-4. In a silken tube covered with frass and grass particles, on the roots and stem-bases of grasses.
P. 5-6. In the larval gallery.
I. 6-7. Frequents coastal sandhills and heaths; easily disturbed in the afternoon and flies at night.

1270(67). **C. nemorella** (Hübn.) (*pratellus* sensu auctt.)
O. On grasses, especially *Deschampsia cespitosa*.
L. 7-4. On the base of the stems, from a silken gallery.
P. 4. In a firm silken cocoon just below the surface of the soil.
I. 5-8. Comes to light.

1271(82). **C. perlella** (Scop.)
O. On grasses, especially *Deschampsia* and *Festuca* spp.
L. 9-6. In a silken gallery at the base of the stems.
P. 6-7. In a silken cocoon on or partly in the ground.
I. 7-8. Comes freely to light.

1272(91). **Agriphila selasella** (Hübn.)
O. On *Puccinellia maritima, Festuca ovina, Spartina maritima* or other grasses.
L. 3-5. In a tubular silken grass-covered gallery on the ground amongst the stems.
P.
I. 7-8. Frequents borders of salt-marshes and fens; easily disturbed in the daytime, flies at dusk and later comes to light.

1273(68). **A. straminella** ([D. & S.]) (*culmella* sensu auctt.)
O. On *Festuca ovina* or other grasses.
L. 9-6. In a vertical silken gallery along the lower part of a stem.
P. 6. In a silken frass-covered cocoon.
I. 6-8. Flies at dusk and later comes to light.

1274(90). **A. tristella** ([D. & S.])
O. On *Deschampsia cespitosa, Poa* spp. or other grasses.
L. 9-6. In a vertical silken gallery along the lower part of a stem but not reaching the ground.
P. 6-7. In a silken cocoon in loose soil amongst the grass-roots.
I. 7-9. Flies at dusk and later comes to light.

1275(84). **A. inquinatella** ([D. & S.])
O. On *Festuca ovina, Poa* spp., or other grasses.
L. 3-5. On the roots and stem-bases, from a slight silken gallery on the surface of the ground.
P. 5-6.
I. 7-8. Frequents dry grassland; flies at dusk and later comes to light and sugar.

1276(81). **A. latistria** (Haw.)
O. On *Bromus* spp. or other grasses.
L. In a silken gallery on the surface of the soil.
P. In a long, narrow, silken sand-covered cocoon.
I. 7. Frequents sandhills and sandy heaths; rests amongst gorse or heather during daytime.

1277(88). **A. poliellus** (Treits.)
O. *On grasses, including *Poa annua*.
L. *Feeds in the autumn; overwinters in a cocoon.
P. *Pupates after hibernation.
I. 9. One specimen taken at Deal in 1885.

1278(85). **A. geniculea** (Haw.)
O. On grasses.
L. 9-6. In a tubular silken tunnel amongst the stem-bases.
P.
I. 7-9. Comes readily to light.

1279(77). **Catoptria permutatella** (Herr.-Schäff.) (*myella* sensu auctt.)
O.
L. On mosses.
P.
I.7-early 8. Often rests in the day time in small pines, preferring isolated trees; flies at dusk.

1280(–). **C. osthelderi** (de Lattin) *Ent.Rec.* **75**:141(1963)
O.
L. *On moss.
P.
I. 7. One specimen taken at light at Bexley Heath, Kent in 1962.

1281(–). **C. speculalis** (Hübn.) *Ent.Gaz.* **18**:91-94(1967)
O.
L.
P.
I. 7. A specimen was taken in Perthshire in 1890; a second, without data, may be from the same locality. The species occurs in the mountains of central and southern Europe.

1282(78). **C. pinella** (Linn.)
O. On *Deschampsia cespitosa, Eriophorum vaginatum* and other grasses or sedges.
L. 9-6. In a vertical silken gallery.
P. 6-7. In a white silken cocoon covered with particles of gnawed grass, in a grass-sheath.
I. 7-8. Often rests in the daytime in small pines; flies at night and comes to light.

1283(79). **C. margaritella** ([D. & S.])
O.
L. On mosses.
P.
I. 7-8. Frequents damp heaths and moors; flies at dusk and after dark.

1284(80). **C. furcatellus** (Zett.)
O.
L. Amongst *Lycopodium* spp.
P.
I. 7-8. Frequents high ground from 450 to 900 metres.

1285(75). **C. falsella** ([D. & S.])
O. On *Tortula* spp., or other mosses growing on old buildings.
L. 2-5. In a silk-lined gallery running through the tufts of the foodplant.
P. 5-7. In the larval gallery.
I. 7-8. Rests by day on moss-covered walls or in thatch or amongst dense foliage; flies at dusk and later comes to light.

1286(76). **C. verellus** (Zinck.)
O.
L. 3-5. On mosses growing on the trunks of trees, especially *Malus, Prunus* or *Populus*.

P.
I. 7-8. Frequents mature orchards.

1287(89). **C. lythargyrella** (Hübn.) (*lithargyrellus* missp.)
O.
L. *On *Poa* spp., *Lolium* spp or mosses; overwinters in a cocoon in the ground.
P.
I. 8-9. Five specimens taken at Deal in 1889.

1288(71). **Chrysocrambus linetella** (Fabr.)
O.
L.
P.
I. 6. Rare immigrant.
(73). *C. dentuellus* (P. & M.) ((72). *cornutellus* (P. & M.)) was erroneously placed on the British list.

1289(70,71). **C. craterella** (Scop.) (*rorella* Linn.)
O.
L. *On *Festuca* and other grasses.
P.
I. 6. An occasional immigrant to the coast of south-east England.

1290(74). **Thisanotia chrysonuchella** (Scop.)
O. On *Festuca ovina, Poa* spp. or other grasses.
L. 7-5. On the stem-bases from a silken tube.
P. 5. In a cocoon in the larval tube.
I. 5-6. Frequents chiefly chalk downs; flies at dusk and after dark.

1291(83). **Pediasia fascelinella** (Hübn.)
O. On *Agropyron* spp., *Deschampsia* spp. or other grasses.
L. 9-6. In a subterranean tube composed of grains of sand spun together with silk.
P. 6-7. In a long, silken, sand-covered cocoon spun vertically at the mouth of the larval tube; the cocoons are sometimes exposed by the wind.
I. 7. Frequents coastal sandhills; seldom disturbed by day but flies freely after dark and comes to light.

1292(86). **P. contaminella** (Hübn.)
O. On *Festuca ovina* or other grasses; hatches in 14-21 days.
L. 9-6. In a funnel-shaped vertical tube amongst the grass blades.
P. 6-7. In a cocoon in the earth.
I. 7-8. Flies only after dark, when it comes to light; cannot easily be disturbed in the daytime.

1293(87). **P. aridella** (Thunb.) (*salinellus* (Tutt))
O. On *Puccinellia maritima, P. fasciculata* or other grasses.
L. 9-5. In a silken, frass-covered, tubular gallery along the ground, attached to the roots or stem-bases of the foodplant.
P. 6. In a cocoon in the earth.
I. 6-8. Frequents the margins and drier parts of salt-marshes; flies after dusk and comes to light; cannot easily be disturbed in the daytime.

1294(93). **Platytes alpinella** (Hübn.)
O.
L. On *Tortula* spp. and other mosses.
P.
I. 7-8. Frequents sandy coasts; flies at dusk and after dark when it visits ragwort blossom and comes to light.

1295(94). **P. cerussella** ([D. & S.])
O. On *Festuca* spp. or other grasses; perhaps also on *Carex* spp.
L. 4-5. On the roots, preferring those growing under stones in sandy soil.
P.
I. 6-7. Frequents sandy coasts and heaths and, less often, chalk hills; flies at dusk and later comes to light; flies again at sunrise.

1296(97A). **Ancylolomia tentaculella** (Hübn.) *Ent.Rec.***64**:273(1952) and **65**:148(1953)
O. *On grasses, especially *Dactylis*.
L. *9-6. In a silken tube spun vertically at the base of a stalk.
P.
I. A rare immigrant to the coast of south-east England; flies at dusk and later comes to light.

1297(99). **Schoenobius gigantella** ([D. & S.])
O. On *Phragmites* or *Glyceria maxima*.
L. 5-6. In the young shoots, often below the surface of the water; when it has finished feeding in a stem it cuts off a piece of reed about its own length and floats on this on the surface of the water until it reaches a new stem.
P. 6-7. In a stem below a prepared exit-hole.
I. 7. Flies at dusk and later comes to light; female restless when boxed.

1298(100). **S. forficella** (Thunb.)
O. On *Glyceria maxima* or *Phragmites*.
L. 5-6. In the young shoots, which it rolls together; it moves to a fresh plant in the same manner as the previous species.
P. 6. In the larval habitation.
I. Frequents ditches, marshes and fens; flies at dusk and later comes to light.

1299(101). **Donacaula mucronellus** ([D. & S.])
O. On *Carex riparia, Phragmites* or *Glyceria maxima*; possibly on other *Carex* spp.
L. 7;9-5. In a stem just above the root.
P. 5-6;7-8. In a stem; a small, round exit-hole 50-75 mm above the root indicates its presence.
I. 6-7;8. Frequents fens, marshes and ditches; flies from 1800 to 2000 hrs and later comes to light. The second brood appears to be a partial one. Female restless when boxed.

1300(98). **Acentria nivea** (Ol.)
O. 6-9. Laid in batches on living or dead parts of *Elodea, Potamogeton* spp., *Chara* spp. or other water-plants, hatching in about 14 days.

L. 9-7. In loosely spun leaves below the surface of the water at depths up to 50 mm.
P. 6-9. In a cocoon attached to a water-plant below the surface of the water.
I. 6-9. Long emergence period, but individuals live only two days. Rests by day at the base of water-plants or on floating material. Males and winged females come readily to light; wingless females normally stay below the surface of the water.

1301(160). **Scoparia subfusca** Haw. (*cembrae* Haw.)
O. On *Picris hieracioides* or *Tussilago.*
L. 9-5. On the roots, in a slight web.
P. 5-6. In a cocoon, usually spun next to the root on which the larva has fed.
I. 6-8. Comes to light.

1302(161). **S. pyralella** ([D. & S.]) (*arundinata* (Thunb.); *dubitalis* (Hübn.))
O.
L. 4-5. On decaying vegetable matter, in a slight web; probably also on the roots of *Senecio jacobaea* and other plants.
P.
I. 6. Flies freely at dusk; easily disturbed in the daytime.

1303(163). **S. ambigualis** (Treits.)
O.
L. Possibly on mosses.
P.
I. 5-6. Rests on tree-trunks and fences in the daytime; flies at dusk and later comes to light.

1304(162). **S. basistrigalis** Knaggs
O.
L.
P.
I. 7. Rests on tree-trunks in the daytime; flies at night and comes to light.

1305(164). **S. ulmella** Knaggs
O.
L.
P.
I. 7-early 8. Rests by day on the trunks of elm and oak, particularly those covered with lichen.

1306(159). **Eudonia pallida** (Curt.)
O.
L. Probably on mosses and lichens growing on the ground.
P.
I. 6-7. Frequents marshes and fens; flies at dusk and later comes to light.

1307(153). **E. alpina** Curt.
O.
L.
P.

I. 6-7. On mountains above 900 metres in the Scottish Highlands, but at sea-level in the Orkneys and Shetlands; flies in afternoon sunshine.

1308(157). E. **crataegella** (Hübn.)
O.
L. 3-5. In a silken tube on mosses, preferring those growing on tree-trunks.
P. 5-7. In the moss.
I. 6-7. Rests on tree-trunks in the daytime, preferring those in open places.

1309(155). E. **murana** (Curt.)
O. Laid singly or in groups of two to five, thrust down into mosses such as *Hypnum cupressiforme*, *Dicranum scoparium*, *Bryum capillare* or *Grimmia pulvinata*.
L. 2-5;7. On mosses growing on old walls and rocks.
P. 6;7. In a cavity under the moss.
I. 6-7;8. Rests on rocks in the daytime; flies at night and comes to light.

1310(158). E. **truncicolella** (Staint.)
O.
L. 9-6. In a silken tube, on mosses growing on the ground.
P. 6-7. In the moss.
I. 7-8. Rests in the daytime on tree-trunks, flying off as one approaches the tree; prefers pine-woods. Flies at dusk and after dark visits flowers, especially field scabious.

1311(151). E. **lineola** (Curt.)
O.
L. 3-6. On the lichens *Parmelia parietina* and *P. olivacea* growing on fences and branches of hawthorn and blackthorn, and on old walls and rocks.
P. 6-7. In a slight cocoon under the lichen.
I. 7-8. Rests in the daytime on lichen-covered branches or fences; flies at dusk and later comes to light.

1312(152;153). E. **angustea** (Curt.) (*vafra* Meyr.)
O. On mosses such as *Tortula muralis* and *Barbula revoluta*, especially those growing on old walls.
L. 4-8. In a silken tube running through the foodplant.
P. 5-8. In a slight cocoon in the moss.
I. 5-10. Adults appearing in the spring may have overwintered. Rests by day on mossy walls and fences; flies at dusk and later comes to light and ivy-blossom.

1313(150). E. **vandaliella** (Herr.-Schäff.) (*resinella* sensu auctt.; *resinea* sensu auctt.)
O.
L. 5. On lichens and mosses growing on the trunks of ash, apple and elm.
P. 6.
I. 7-8. Rests by day on the trunks bearing its foodplant; flies in the evening and later comes to light and sugar.

1314(156). **E. mercurella** (Linn.) (*frequentella* (Staint.))
O. On various mosses growing on tree-trunks, rocks and stone walls.
L. ?9-4. In a silken gallery in the moss, probably from the autumn.
P. 5-6. In the moss.
I. 6-8. Rests on tree-trunks, rocks and walls in the daytime.

1315(105). **Nymphula nymphaeata** (Linn.)
O. 7-8. Laid in a flat mass on the underside of a leaf of *Potamogeton* spp., *Hydrocharis, Sparganium* spp. or some other aquatic plant, hatching in about 14 days.
L. 7-6. First mining the underside of a leaf; later in a flat, oval case made of pieces of a leaf; when feeding, it attaches the case to the under surface of a leaf with silk; otherwise the case floats freely on the surface of the water.
P. 6-7. In a cocoon covered with pieces of leaf, attached to the stem of a plant about 25 mm above water level.
I. 6-8. Flies at dusk and later comes to light.

1315A (–). N. **enixalis** Swinhoe *Ent.Gaz.***29**:118-119(1978)
O. On *Potamogeton* and various other aquatic plants.
L. When young, in a tight spinning often below the surface of the water; later in a flat case of leaf-fragments, which floats to the surface.
P.
I. Rests head-downwards. A far-eastern species introduced into aquatic nurseries, where it is continuous-brooded.

1315B(–). N. **melagynalis** Agassiz *Ent.Gaz.***29**:121-123(1978)
O.
L.
P.
I. Rests head-downwards. A tropical species accidentally introduced into aquatic nurseries.

1316(104). **Parapoynx stratiotata** (Linn.)
O. On *Potamogeton* spp., *Elodea, Ceratophyllum demersum* or some other aquatic plant.
L. 7-5. Amongst spun leaves, below the surface of the water.
P. 6. In a large, pinkish, oval cocoon attached to a stem or debris.
I. 6-8. Flies at dusk and later comes to light.

1317(–). P. **obscuralis** (Grote) *Ent.Gaz.***19**:107-112(1968)
O.
L. On aquatic plants below the surface.
P.
I. Occurs in heated greenhouses amongst water-plants imported from the U.S.A.; more or less continuously brooded.

1318(103). P. **stagnata** (Don.)
O. On *Sparganium* spp. or *Nuphar lutea*.
L. 8-5. Before hibernation, mining the leaves or stem; later, amongst spun leaves just below the surface of the water.
P. 6-7. In a white silk cocoon attached to the foodplant at or just below water-level.
I. 7-8. Flies at dusk and later comes to light.

1318A (–). **P. diminutalis** Snellen *Ent.Gaz.* **29**:123-125(1978)
O.
L. On various aquatic plants, often feeding totally submerged.
P.
I. A far-eastern species introduced into aquatic nurseries.

1318B(–). **Oligostigma angulipennis** Hampson *Ent.Gaz.* **29**:125-126(1978)
O. On *Cryptocoryne* spp.
L. In a wet spinning between two leaves.
P. In the larval spinning.
I. A native of South India and Sri Lanka, introduced into aquatic nurseries.

1318C(–). **O. bilinealis** Snellen *Ent.Gaz.* **29**:126-127(1978)
O.
L.
P.
I. A far-eastern species, a single specimen of which was found in an aquatic nursery in Hampshire in 1977.

1318D(–). **O. polydectalis** (Walker)
O.
L.
P.
I. An exotic species, a single specimen of which was taken in an aquatic nursery in 1978 and three more in 1979.

1319(102). **Cataclysta lemnata** (Linn.)
O. 7-8. Laid in small batches on a leaf of *Lemna* spp. on or under the surface of the water; hatch in 10-12 days.
L. 8-5. In an irregular, oval, portable case constructed of leaf-fragments of the foodplant; the larva feeds below the surface of the water, but its case contains air.
P. 5-6. In a dense oval cocoon of white silk covered with leaf fragments attached just below the surface of the water.
I. 6-8. Males fly at dusk over the surface of the water; females fly at night. Both sexes come to light.

1320(–). **Synclita obliteralis** (Walk.) *Ent.Gaz.* **19**:155-158(1968)
O.
L. On aquatic plants.
P.
I. A North American species which was accidentally introduced with water-plants to heated greenhouse nurseries; more or less continuously brooded.

1321(167). **Evergestis forficalis** (Linn.)
O. Laid in batches, usually on the underside of a leaf, on *Brassica* spp., *Raphanus* spp. or some other species of crucifer.
L. 6-7;9-5. In the heart or amongst young spun leaves of the foodplant. Autumn larvae overwinter in their cocoons.
P. 5;7. In a cocoon beneath the surface of the soil.
I. 5-6;8-9. Flies at dusk and later comes to light. Sometimes a pest on garden cabbages.

1322(166). **E. extimalis** (Scop.)

O. On *Sinapis, Iberis, Diplotaxis tenuifolia, Sisymbrium* spp. or some other species of crucifer.

L. 8-5. On the seed-pods in a silken web which may contain more than one larva; hibernates in its cocoon.

P. 5. In a tough cocoon in the soil.

I. 6-7. May be bivoltine. Flies at dusk and later comes to light. Easily disturbed by day. A migrant which establishes colonies which persist for several years.

1323(165). **E. pallidata** (Hufn.) (*straminalis* sensu Hübn.)

O. 8. On *Barbarea vulgaris* or some other species of crucifer; hatches in about 14 days.

L. 8-5. On the leaves gregariously, the larvae resting side by side on a leaf when not feeding; they hibernate in their cocoons.

P. 5-6. In a cocoon under leaves on the surface of the ground.

I. 6-8. Flies at dusk and later comes to light; not easily disturbed in daytime.

1324(149). **Cynaeda dentalis** ([D. & S.])

O. On *Echium vulgare* or, less often, *Anchusa officinalis*.

L. ?8-6. In the lower part of the stem and leaf-stalks, aborting them.

P. 6-7. In a strong, hard cocoon covered with pieces of leaf, spun on a withered basal leaf; found by feeling round the base of the plants.

I. 7. Flies at dusk and later rests on the flowers of the foodplant.

1325(–). **Hellula undalis** (Fabr.) *Ent.Gaz.* **19**:111-112(1968)

O.

L. *On Cruciferae.

P.

I. 9. A single specimen was taken in south Devon in 1967. The species has a wide distribution in Europe, Asia, Africa and the Pacific Islands.

1326(133). **Pyrausta aurata** (Scop.)

O. On *Nepeta, Origanum, Mentha* spp., *Salvia* spp. or *Calamintha* spp.

L. 6-7;9-4. When young, on the under surface of a leaf; later on the flowers in a web. Autumn larvae overwinter in their cocoons.

P. 4;7. In a cocoon on a flower-head.

I. 5-6;7-8. Flies in sunshine and after dark, when it comes to light. Sometimes a garden pest.

1327(130). **P. purpuralis** (Linn.)

O. On *Mentha arvensis* or *Thymus drucei*.

L. 6;7-9. Between spun leaves.

P.

I. 5-6;7-8. Flies in sunshine and after dark, when it comes to light.

1328(132). **P. ostrinalis** (Hübn.)

O.

L.

P.

I. The differences in life histories between this species and *P. purpuralis* have not yet been studied.

1329(131). **P. sanguinalis** (Linn.)
O. On *Thymus drucei*.
L. 7;9-4. On the flowers, in a silken tube. Autumn larvae overwinter in their cocoons.
P. 5;7-8. In a tough cocoon amongst moss on the ground and usually attached to some solid object.
I. 6;8. Flies in sunshine.

1330(134). **P. cespitalis** ([D. & S.])
O. On *Plantago lanceolata* or *P. major*.
L. 6;8-9. Gregarious in silken galleries spun close to the base of a leaf, feeding nocturnally on the underside of a leaf.
P. 4;6-7. In the web.
I. 5;7-8. Flies in sunshine and at dusk; later comes to light.

1331(129). **P. nigrata** (Scop.)
O. On *Origanum, Thymus drucei, Mentha arvensis* or *Galium odoratum*; hatching in about 14 days.
L. 6-7;9-10. In a slight web on the underside of the leaves, making at first white blotches and later holes.
P. 7-8;11-5. In an oval, whitish cocoon.
I. 5-6;8. Flies in sunshine.

1332(128). **P. cingulata** (Linn.)
O. On *Salvia pratensis* or *Thymus drucei*.
L. 6;8-4. Under a slight web on the underside of the leaves, preferring those resting on the ground; autumn larvae hibernate in their cocoons.
P. 4;6-7. In a rather large, greyish brown, papery cocoon.
I. 5-6;7-8. Flies in sunshine.

1333(146). **Margaritia sticticalis** (Linn.)
O. 6;8. On *Artemisia vulgaris* or *A. campestris*. hatching in a few days.
L. 6-7;9. On the upperside of a leaf, not eating the lower substance.
P. 7;7-6;9-6. Overwinters as a pupa, second generation from September.
I. 6;7-8. Only part of the spring generation emerges in July and August; comes to light.

1334(145). **Uresiphita limbalis** ([D. & S.]) (*polygonalis* sensu auctt.)
O. *On *Genista* or *Sarothamnus*.
L. *7-8.
P.
I. A casual immigrant.

1335(148). **Sitochroa palealis** ([D. & S.])
O. On *Daucus, Pastinaca, Heracleum sphondylium* or *Foeniculum*.
L. 8-5. On the flowers and seeds, compressing them together with silken threads; hibernates from September, sometimes for two winters, in its cocoon.
P. 5. In a tough silken cocoon in the earth.
I. 6-7. An immigrant which often establishes itself temporarily; flies at dusk and later comes to light; rests on flowers by day, when it is easily disturbed.

1336(147). **S. verticalis** (Linn.)

O.
L. *7-5. On *Cirsium arvense, Teucrium* spp., *Sarothamnus, Atriplex* spp., *Rumex* spp. and *Diplotaxis tenuifolia*; overwinters in a hibernaculum.
P. *6. In a cocoon amongst spun leaves.
I. 6-7. Readily disturbed by day; later comes to light; population probably reinforced by immigration.

1337(138). **Paracorsia repandalis** ([D. & S.])

O. On *Verbascum nigrum.*
L. 6-7;9-5. In spun shoots and in the flower-heads.
P.
I. 5-7. A casual immigrant which occasionally becomes temporarily established.

1338(144). **Microstega pandalis** (Hübn.)

O. 6. Laid in small batches on leaves of *Teucrium scorodonia, Origanum* or *Solidago virgaurea,* hatching in about 14 days.
L. 7-4. When young, under a slight web on the underside of the leaves; later in a succession of elongate cases made of leaf-fragments and open at both ends. Overwinters full-fed in its case.
P. 5. In the last larval case, closing the ends.
I. 5-6. Flies at dusk.

1339(126). **M. hyalinalis** (Hübn.)

O. 7-8. On *Centaurea nigra* or *Verbascum thapsus.*
L. 8-5. When young, under a slight web on the underside of a leaf, making transparent blotches; after hibernation, on the young leaves from a silken gallery at the base of the foodplant.
P. 6. In a gauzy white cocoon in leaves at the base of the foodplant.
I. 7. Easily disturbed in daytime; flies at dusk and later visits flowers and light.

1340(142). **Ostrinia nubilalis** (Hübn.)

O. Laid in batches on the underside of leaves of *Artemisia vulgaris*, hatching in 7-21 days; has also been recorded on *Malva sylvestris.*
L. 7-4. In a stem; stems containing larvae snap off easily at the entrance hole, generally close to the ground; cf.no.1148.
P. 4-5. In the larval burrow.
I. 5-8. Peak emergence in late June; flies at night and comes to light. On the Continent and in North America a serious pest of maize, and could turn to this foodplant with the increase of its cultivation in Britain.

1341(113). **Eurrhypara hortulata** (Linn.)

O. On *Urtica* or, less frequently, *Marrubium, Stachys* spp., *Mentha* spp. or *Ballota.*
L. 8-5. In spun leaves or a rolled leaf; overwinters in its cocoon.
P. 5. In a cocoon spun under loose bark, deep in debris, etc.
I. 6-7. Flies from dusk onwards and comes to light.

1342(114). **E. lancealis** ([D. & S.])

O. On *Eupatorium, Teucrium scorodonia, Stachys* spp. or *Senecio* spp.

L. 8-5. On the leaves, flowers and seed-heads, living in a leaf spun so as to turn the edges downwards; hibernates in its cocoon.

P. 5. In a white silken cocoon within a pyramidal web.

I. 6-7. Flies at dusk and later comes to light.

1343(123). **E. coronata** (Hufn.) (*sambucalis* ([D. & S.]))

O. On *Sambucus nigra, Viburnum, Syringa* or *Ligustrum*, preferring young plants.

L. 8-5. In a spun fold on the underside of a leaf which it leaves at night in order to feed; hibernates in its cocoon.

P. 5-6. In a cocoon amongst fallen leaves or on bark.

I. 6-7. Flies at dusk and later comes to light; rests by day on fences, being particularly fond of the overlap of the slats.

1344(121). **E. terrealis** (Treits.)

O. Laid in batches on the upperside of leaves of *Solidago virgaurea*, hatching in 14-21 days.

L. 7-5. On the underside of a leaf, or, more often, in a web amongst the blossom, sometimes denuding a plant of its flowers; overwinters in its cocoon.

P. 6-7. In a strong, yellowish white cocoon.

I. 7. Can be disturbed from the foodplant in the daytime; flies at dusk.

1345(137A). **E. perlucidalis** (Hübn.) *Ent.Gaz.* 8:162-166(1957)

O.

L. 6-8. On *Cirsium* spp., especially *C. oleraceum*.

P.

I. 6. Established at Woodwalton Fen, Hunts. and occurs as a scarce immigrant on the south-east coast.

1346(127). **Anania funebris** (Ström)

O. On *Solidago virgaurea* or *Genista tinctoria*.

L. 7-4. Under a slight web on the underside of the lower leaves, but feeding on the flowers as well as the leaves; overwinters in its cocoon.

P. 5. In an oval, yellowish cocoon.

I. 6-7. Flies in sunshine with a characteristic spinning motion.

1347(141). **A. verbascalis** ([D. & S.])

O. On *Teucrium scorodonia* or *Verbascum thapsus*.

L. 8-4. In a web on the underside of a leaf or between joined leaves, eating holes from the underside; overwinters in its cocoon.

P. 5-6.

I. 6-7. Flies at dusk and later comes to light.

1348(125). **A. pulveralis** (Hübn.)

O. On *Mentha aquatica, M. arvensis* or *Lycopus*.

L. 8. On the underside of the leaves.

P.

I. 6-7. An immigrant, occasionally becoming established.

1349(140). **A. stachydalis** (Germ.)
- O. On *Stachys sylvatica* or *S. palustris*.
- L. 8-4. In a web or tube under a leaf, usually low down, the web causing the edges of the leaf to turn down; overwinters in its cocoon.
- P. 5-6. In a cocoon spun in a hollow stem.
- I. 6-7. Flies at dusk and later comes to light; hides in thick vegetation by day, from which it is not easily disturbed.

1350(116). **Ebulea crocealis** (Hübn.)
- O. On *Pulicaria dysenterica* or *Inula conyza*.
- L. 9-5. Hibernates when young in the turned-down edge of a leaf; in the spring in the central shoot which is spun up with silk.
- P. 6. In two or three leaves of the foodplant spun together.
- I. 7-8. Flies at dusk and later comes to light.

1351(122). **Opsibotys fuscalis** ([D. & S.])
- O. On *Rhinanthus*, *Melampyrum pratense*, or *Solidago virgaurea*.
- L. 7-4. In a gallery or web, spinning the flowers and seeds to the main stalk; overwinters in its cocoon.
- P. 5. In a tough semitransparent cocoon.
- I. 6. Easily disturbed in the daytime; flies at dusk and later comes to light.

1352(115). **Nascia cilialis** (Hübn.)
- O. On *Carex riparia* or some other species of *Carex*.
- L. 9-5. Feeds in the evening on the leaves, resting by day outstretched on adjacent herbage; overwinters in its cocoon.
- P. 6. In a papery cocoon in an old reed stem.
- I. 6-7. Flies from sunset onwards and comes to light.

1353(117). **Udea elutalis** ([D. & S.]) (*lutealis* (Hübn.))
- O.
- L. ?9-6. Polyphagous on *Rubus*, *Centaurea*, *Plantago*, *Fragaria*, *Cirsium*, *Artemisia*, etc.; in a web on the underside of, usually, a lower leaf.
- P. 6-7. In a folded leaf or between joined leaves.
- I. 7-9. Easily disturbed in daytime; flies at dusk and later comes to light, flowers and sugar.

1354(119). **U. fulvalis** (Hübn.)
- O. 8. On *Salvia pratensis*, *Nepeta*, *Ballota* or *Lychnis* spp.; hatching in about 14 days.
- L. 8-6. Hibernates without having fed; in the spring feeds on the leaves.
- P. 7. In a cocoon amongst leaves of the foodplant.
- I. 7-8. Comes to light.

1355(120). **U. prunalis** ([D. & S.])
- O.
- L. 10-5. Polyphagous on various Labiatae, other herbaceous plants, bushes and trees; in a slight web on the underside of a leaf; after hibernation, making a rather more substantial spinning.
- P. 6. In a white silken cocoon beneath the edge of a leaf.
- I. 6-7. Conceals itself by day mainly in blackthorn; flies at dusk and later comes to light.

1356(136). **U. decrepitalis** (Herr.-Schäff.)
- O. On *Dryopteris carthusiana* or some other species of fern.
- L. 7-4. In a slight web on the underside of a frond; overwinters in a cocoon.
- P. 5-6. In a strong silken cocoon.
- I. 6-7. Occurs in damp localities in the Scottish Highlands.

1357(137). **U. olivalis** ([D. & S.])
- O.
- L. 9-4. Polyphagous on Labiatae, other herbaceous plants and bushes; in spun leaves or a turned-down edge of a leaf.
- P. 5-6. In a web in a spun leaf.
- I. 6-7. Easily disturbed in daytime; flies from dusk and comes to light.

1358(135). **U. uliginosalis** (Curt.) (*alpinalis* sensu Ford)
- O. On *Senecio jacobaea*, but also occurs where this plant is absent.
- L. 6. Amongst spun leaves.
- P. 6-7.
- I. 7-8. Occurs in the mountains of Scotland above 350 metres. *U. alpinalis* ([D. & S.]) is a distinct species and its occurrence in Britain is not confirmed.

1359(118). **U. ferrugalis** (Hübn.)
- O.
- L. 7-10. Polyphagous on herbaceous plants, notably *Stachys* spp., *Eupatorium* and *Fragaria vesca*; on the underside in a turned-down leaf or between joined leaves.
- P. 8-10. In a cocoon in a partially cut section of leaf; the stage lasts from three weeks to eleven months.
- I. 7;9-10. A common immigrant readily establishing itself. Each stage can occur in almost any month of the year; adults are most often recorded in the autumn, when they visit ripe blackberries, ivy-bloom, light and sugar.

1360(139). **Mecyna flavalis** ([D. & S.])
- O. On *Galium mollugo, Artemisia campestris* or *Urtica urens*.
- L. 3-4.
- P.
- I. 7. A downland species; easily disturbed by day.

1361(143). **M. asinalis** (Hübn.)
- O. On *Rubia*.
- L. 7-8;9-5. In a silken gallery beneath the leaves, making conspicuous white blotches.
- P. 6;8. In a rolled leaf or amongst fallen leaves.
- I. 6-7;8. Flies at dusk and at night and visits flowers.

1362(124). **Nomophila noctuella** ([D. & S.])
- O. 7-8. On *Trifolium* spp., *Polygonum aviculare* or, probably, species of grass; hatches in 14-21 days.
- L. 8-10.
- P. 8 or 8-6. In a tough cocoon amongst the leaves of the foodplant or attached to a stone.

I. 7-8;8-9. A common immigrant establishing itself in favourable seasons. Some adults emerge within a few days of pupation; the remainder the following spring when in captivity, but they seldom, if ever, survive the English winter in the wild.

1363(108). **Dolicharthria punctalis** ([D. & S.])
O. 9. On *Lotus corniculatus*.
L. 9-5. Before hibernation, under a silk web on living or dead leaves of *Lotus*; after hibernation also on adjacent vegetation, preferring decaying leaves.
P. 5-7. On the ground in a tough cocoon mixed with debris.
I. 7-8. Usually near the coast. Often rests in gorse during the daytime; flies at night and comes to light.

1364(109). **Antigastra catalaunalis** (Dup.)
O. *On *Linaria* spp. or *Antirrhinum*.
L. *9-3. Overwinters as larva.
P.
I. 8-9. A rare immigrant.

1365(–). **Maruca testulalis** (Gey.) *Proc.Brit.ent.nat.Hist.Soc.* **1970**:33
O.
L. On *Phaseolus* spp., boring into the pods.
P.
I. Two specimens were bred in 1967 at East Malling Research Station from French beans imported from Malawi.

1366(106). **Diasemia litterata** (Scop.)
O. On *Plantago* or *Hieracium*.
L. 3,4,5,7. In several generations.
P.
I. 6-9. An immigrant which occasionally becomes temporarily established; flies in the evening and later comes to light and sugar.

1367(107). **Diasemiopsis ramburialis** (Dup.)
O.
L.
P.
I. 6-7;9-10. A rare immigrant; flies at night and comes to light and sugar.

1368(167A). **Hymenia recurvalis** (Fabr.) *Ent.Gaz.* **3**:57(1952)
O. *Laid on the underside of a leaf of *Beta* spp., *Spinacia* spp. or *Zea mays*, hatching in about five days.
L. *In a web on the leaves.
P.
I. Specimens were taken in Surrey and Dorset in 1951 and 1976; the early stages have not been recorded in Britain. The species has an almost world-wide distribution.

1369(112). **Pleuroptya ruralis** (Scop.)
O. On *Urtica dioica*; also on *Chenopodium, Atriplex, Filipendula ulmaria* or *Humulus*.
L. ?9-6. In a leaf rolled round the midrib.
P. 6. In a rolled leaf.
I. 7. Flies at dusk and comes to light.

1370(–). **Herpetogramma centrostrigalis** (Steph.)
An American species described from a specimen reputedly taken in Britain.

1371(–). **H. aegrotalis** (Zell.)
One specimen taken in Lancashire in 1889.

1372(110). **Palpita unionalis** (Hübn.)
O. On *Jasminum officinale* in captivity.
L. Continuous brooded at high temperature.
P.
I. 6;8-10. A fairly regular immigrant, most frequent in October; flies at night and comes to light, sugar and flowers such as ivy and *Buddleia*.

1373(–). **P. lucernalis** (Hübn.)
An adventive species figured by Wood, *Index Entomologicus* p.123(1839).

1374(111). **Agrotera nemoralis** (Scop.)
O. 6. On twigs of *Carpinus*, hatching in about ten days; also on *Betula* spp. and *Castanea*.
L. 6-4. When young, on the underside of a leaf under a slight web, later between two spun leaves; overwinters in a cocoon spun in a partially cut and folded section of a leaf.
P. 4-6. In the cocoon which falls from the tree in winter.
I. 5-7. Flies at dusk and later comes to light. Adults occasionally emerge in August.

1375(–). **Leucinodes vagans** Tutt
An African species described in Britain and, therefore, erroneously supposed to have been taken in this country.

1376(–). **Daraba laisalis** (Walker)
An African species, of which one specimen was taken at light near Heathrow airport in 1973.

1377(170). **Hypsopygia costalis** (Fabr.)
O.
L. 9-5. On clover-hay and grass-hay; in squirrels' dreys; probably also in thatch and dead leaves.
P. 6-7. In an oval cocoon in the feeding place.
I. 7-8. Sometimes abundant round old haystacks; flies at night and comes to light and sugar.

1378(177). **Synaphe punctalis** (Fabr.) (*angustalis* ([D. & S.]))
O. On *Hypnum cupressiforme*, or, possibly, other species of moss growing on the ground in damp places.
L. ?9-6. In a slight web concealed in the moss, but ejecting frass on to the surface.
P. 6. In a silken cocoon amongst the moss.
I. 6-8. Easily disturbed in the daytime; flies in the evening and later the male comes to light.

1379(169). **Orthopygia glaucinalis** (Linn.)
O.
L. 3-5. On dead and decaying vegetable matter of all types; found

in haystacks, thatch, witches'-brooms, birds' nests, squirrels' dreys, piles of leaves, etc., living in silken galleries.
P. 5-6. In a cocoon in the feeding place.
I. 7-8. Flies at dusk and later comes to light and sugar. Sometimes very common in old stack-yards.

1380(172). **Pyralis lienigialis** (Zell.)
O.
L. Possibly in vegetable refuse or in stored products.
P.
I. 6;7-9. Has been found resting on walls; flies at night and comes to light.

1381(171). **P. farinalis** (Linn.)
O. Laid loose, hatching in three to five days.
L. 9-5. On stored cereals and cereal refuse, in a long silken tube; sometimes living for two years.
P. 5-7. In a tough cocoon in the feeding place.
I. 6-8. Sometimes plentiful in old stables, barns, warehouses and mills; flies at dusk and later comes to light and sugar.

1382(–). **P. manihotalis** Guen.
O.
L. On skins, bones and hides.
P.
I. Found in 1943 in material imported from India.

1383(173). **P. pictalis** (Curt.)
O.
L. On stored grain.
P.
I. A single specimen taken in London, no doubt imported.

1384(175). **Aglossa caprealis** (Hübn.) (*cuprealis* missp.)
O. Hatch in 14-21 days.
L. 8-5. In a silken tube amongst chaff and hay-refuse in stables and barns, often living nearly two years. In confinement will eat grass-seed.
P. 6-7. In a silken cocoon to which straw and refuse are attached.
I. 7-8. Rests on the walls in daytime.

1385(174). **A. pinguinalis** (Linn.)
O.
L. 8-5. In a silken tube amongst chaff and refuse of cereals, seeds, hay and sheep dung in stables, barns and warehouses; often living nearly two years.
P. 5-6. In a grey silken cocoon covered with gnawed particles, constructed some distance from the feeding place.
I. 6-7. Rests by day on the walls in dark corners; flies at dusk and later comes to light. Formerly common in stables.

1386(176). **A. dimidiata** (Haw.)
O.
L. On stored tea and rice and on dried insects.
P.
I. Has been taken in tea warehouses.

1387(176A). **A. ocellalis** Led.
Found in Glasgow in 1943 in a cargo of West African palm kernels.

1388(168). **Endotricha flammealis** ([D. & S.])
O. 7-8. Laid on the axils of the flowers and stalks of *Lotus uliginosus*; probably also on other foodplants, including *Corylus, Quercus* spp., *Salix* spp., *Carpinus, Ligustrum* and *Agrimonia eupatoria*; hatching in 10-14 days.
L. 8-5. When young, in a chamber divided into 3-5 compartments spun on the underside of a leaf; later on decaying leaves on the ground, feeding in mild weather throughout the winter.
P. 5-6. In a cocoon attached to a dead leaf on the ground.
I. 6-8. Flies at dusk and later comes to light and sugar; easily disturbed in the daytime.

1389(60). **Galleria mellonella** (Linn.)
O. Laid in rows in a crevice; hatch in 9-30 days.
L. 8-5. On the comb in bee-hives, preferably an old comb, in a silk-lined gallery; overwinters in its cocoon.
P. 6-7. In a cocoon spun in the comb or hive, sometimes gregariously.
I. 7-10. Often rests in the daytime under the hive; flies at night and sometimes comes to light or sugar.

1390(55). **Achroia grisella** (Fabr.)
O.
L. 9-5. On wax in bee-hives, preferring old combs; also on dried fruit and dead insects.
P. 6-7. In a white cocoon, usually near the entrance to the hive.
I. 6-10. There may be two overlapping broods. Flies round the hives at dusk.

1391(56). **Corcyra cephalonica** (Staint.)
O. Laid singly or in small batches on the food; hatching in about seven days.
L. 8;11-5. On cacao, dried currants and other dried fruit, and on various kinds of stored grain and foodstuff.
P. 6-7;8-9. In a thick white cocoon, sometimes incorporating frass and debris, spun near the food; sometimes gregarious.
I. 7;9-10. Sits on walls in the daytime; flies at night.

1392(58). **Aphomia sociella** (Linn.)
O.
L. 8-4. On the comb in the nests of wasps and humble-bees in a silken tube or gallery; not often in subterranean nests. Overwinters in its cocoon.
P. 4-5. In a cocoon in the nest, usually spun gregariously; sometimes outside the nest.
I. 6-8. Flies at dusk and later comes to light.

1393(57). **Melissoblaptes zelleri** (Joannis) *Ent.Rec.* **48**:93-94(1936)
O. 7. Hatch in about three weeks.
L. 8-6. On the moss *Brachythecium albicans*; in a vertical silken tube extending about 100 mm into the sand, from the upper end of which one to three other tubes radiate on the

surface towards its foodplant.
P. 6-7. In the silken tube below the surface.
I. 7-8. Males rest on the ground or herbage after dark, vibrating their wings; females make short flights. Attracted by light, but does not enter a trap. Flies freely just after dark and again in the early morning sun; sometimes rests on *Echium* at night.

1394(59). **Paralipsa gularis** (Zell.)
O. Laid in batches on or near the food, hatching in one to three weeks.
L. 5-3. On stored almonds, walnuts, grain, and dried fruit; overwinters in a cocoon spun in a crevice often at a distance from the foodstuff.
P. 3-5. In a tough cocoon, often gregariously, in crevices high on warehouse walls.
I. 5-8. In warehouses, in overlapping broods.

1395(–). **Arenipses sabella** Hamps.
O.
L. On stored dates.
P.
I. Has been accidentally imported into Britain in cargoes of dates.

1396(1). **Anerastia lotella** (Hübn.)
O.
L. ?9-6. Probably from the autumn, in a tubular case covered with sand, on stems and the crown of roots of *Ammophila, Corynephorus, Festuca ovina* and probably other grasses growing on sandy coasts and localities inland.
P. 6-7. In a cocoon in sand at the base of the foodplant.
I. 7. Flies at dusk and later comes to light; in the daytime, rests on grass stems and may be taken by tapping them and watching for the moth to drop to the sand; at night, moths resting near the roots of marram are conspicuous by lamplight.

1397(50). **Cryptoblabes bistriga** (Haw.)
O. On *Quercus* spp. or, less often, *Alnus glutinosa*.
L. 8-10. In a folded leaf, eating out holes between the veins.
P. 11-6. In detritus on the ground.
I. 6-7. Flies at dusk and later comes to light.

1398(30). **C. gnidiella** (Mill.)
O.
L. 8-11. In pomegranates, in a conspicuous web in the top of the fruit; on oranges.
P. 11. In the feeding place.
I. Present amongst pomegranates in most greengrocers' shops; will feed up slowly if kept in the warmth and the adults will emerge in late winter. Unlikely to survive in the open.

1399(54). **Acrobasis tumidana** ([D. & S.])
O. On *Quercus* spp.
L. 9-6. In spun leaves.
P. 6-7.
I. 7-8. Flies at dusk and later comes to light and sugar.

1400(53). **A. repandana** (Fabr.) (*tumidella* (Zinck.); ?*cristella* sensu Beirne)
O. On *Quercus* spp.
L. 5-6. In spun leaves, generally high in the tree.
P. 6-7.
I. 7-8. Flies at dusk and later comes to light and sugar.

1401(51,52). **A. consociella** (Hübn.) (*sodalella* sensu auctt.)
O. On *Quercus* spp., preferring saplings.
L. 9-6. Gregariously in a conspicuous web spun over the leaves.
P. 6-7. In the feeding place.
I. 7-8. Flies at dusk and later comes to light and sugar.

1402(49). **Eurhodope suavella** (Zinck.)
O. On *Prunus spinosa*, or, less often, *Crataegus*, or *Sorbus* spp., preferring stunted and isolated bushes.
L. 9-6. In a thick silken tube under the leaves and on the twigs, covered with frass and fragments of dead leaf.
P. 6. In the tube.
I. 7. Flies at night and comes to light.

1403(48). **E. advenella** (Zinck.)
O. On *Crataegus*, or, less often, *Sorbus aucuparia*.
L. 5-6. In spun flowers and young leaves.
P. 6-7. In earth or in leaf-litter.
I. 7-8. Flies at night and comes to light.

1404(47). **E. marmorea** (Haw.)
O. On *Prunus spinosa*, preferring stunted bushes; on *Crataegus* or *Sorbus* spp.
L. 5. In a slight silken tube spun on the underside of a twig, leaving the tube to feed; the tube resembles a small piece of sheep's wool.
P. 5-6. In the web.
I. 6-7. Flies at dusk and later comes to light.

1405(7). **Oncocera semirubella** (Scop.)
O. On *Lotus corniculatus, Trifolium repens, Ononis, Hippocrepis* or *Medicago* spp.
L. 8-6. In a dense web, sometimes gregariously.
P. 6-7. In a dense cocoon amongst the foodplant, on the surface of the ground or in spun leaves.
I. 7. Easily disturbed in sunny weather; flies just before dusk and later comes to light and sugar.

1406(10). **O. palumbella** ([D. & S.])
O. On *Erica, Calluna, Polygala* spp. or *Thymus drucei*.
L. 7-6. In a tubular, frass-covered web.
P. 6-7. In a cocoon in earth or at the base of the foodplant.
I. 7-8. Easily disturbed in the daytime; flies at dusk and after dark.

1407(20). **O. genistella** (Dup.)
O. 8. On *Ulex*, preferring *Ulex minor*, on the young shoots.
L. 9-6. In a web more or less gregariously; in fine weather suns itself on the outside of the web.
P. 6-7. In the web.
I. 7. Can be beaten from the foodplant in the daytime.

1408(11). **O. obductella** (Zell.) *Ent.* **62**:193(1929)
O. 8. On the flower-heads or upper leaves of *Origanum*; on *Mentha arvensis* or *Calamintha* spp.
L. ?9-7. In a spun terminal shoot, making a large web intermixed with dead leaves of the plant.
P. 7. In earth or amongst leaf-litter.
I. 7-8. Comes to light; can be disturbed in the early evening.

1409(5). **O. formosa** (Haw.)
O. 7. On *Ulmus* spp., preferring bushes.
L. 7-9. Feeds openly by day from a scanty white web spun along the midrib on the upperside of a leaf.
P. 9-6. In earth or in rotten wood.
I. 6-7. Flies at dusk and later comes to light.

1410(–). **Salebriopsis albicilla** (Herr.-Schäff.) *Ent.Gaz.* **16**:13-14(1965)
O. On *Tilia cordata*; *also on *Betula* spp., *Corylus* or *Alnus glutinosa*.
L. 8-9. In a rolled leaf.
P.
I. 6. Comes to light.

1411(21,6). **Nephopterix hostilis** (Steph.) (*adelphella* sensu auctt.)
O. On a leaf of *Populus tremula*.
L. 7-9. When young, in a rolled leaf or turned-down edge; later in a chamber formed by spinning a dead or dying leaf to a living leaf, with strands of silk extending to other leaves, more than one larva sometimes being found in one chamber.
P. 8-6. In a tough cocoon spun amongst leaf-litter.
I. 6. Flies at night and comes to light.

1412(–). **Selagia argyrella** ([D. & S.])
O. *On *Calluna*.
L. *9-6. In a web; overwinters in a silken tube.
P. *In earth.
I.

1413(22). **Microthrix similella** (Zinck.)
O. On *Quercus* spp., probably near the top of the tree.
L. 7-8. In a silk web on the leaves.
P. 4-5. Has been reared from a pupa found in a slight web in a crevice of bark at the base of an oak-tree.
I. 6. Comes to light at about 0130 hrs; otherwise seldom seen owing to its habit of keeping to the tree-canopy.

1414(9). **Metriostola betulae** (Goeze)
O. On *Betula* spp.
L. 5-6. Under a silk web on the upperside of a leaf.

P. 6-7. In a cocoon amongst spun leaves or in leaf-litter.
I. 7. Flies at dusk and later comes to light.

1415(8). **Pyla fusca** (Haw., 1811)
O. On *Erica* or *Vaccinium* spp.; hatches in about 14 days.
L. 8-5. In a web.
P.
I. 6-7. In the daytime, rests especially on burnt parts of a heath and flies vigorously when disturbed; flies at night and comes to flowers and light.

1416(19). **Phycita roborella** ([D. & S.]) (*spissicella* (Fabr.))
O. On *Quercus* spp.; also on *Pyrus* or *Malus*.
L. 9-5. In spun leaves; can be obtained by searching or beating.
P. 6-7. In leaf-litter or in the earth.
I. 7-8. Can be disturbed from the lower branches in the daytime; flies at dusk and later comes to light, flowers and sugar.

1417(13). **Pima boisduvaliella** (Guen.)
O. On *Lathyrus japonicus, Anthyllis, Lotus corniculatus* or *Ononis spinosa*.
L. 7-5. In a pod. Overwinters full-fed in a spherical hibernaculum of silk mixed with sand.
P. 5-7. In an oval, sand-covered cocoon on or in the ground.
I. 6-8. Flies in evening sunshine.

1418(17). **Dioryctria abietella** ([D. & S.])
O. On *Pinus sylvestris*.
L. 8-5. In a cone or shoot exuding much frass, but not causing resinous exudation; when feeding in a cone, overwinters in a circular hibernaculum on its surface which it leaves in the spring to pupate.
P. 6-7. In the mine, in rotten wood or detritus.
I. 7-8. Best obtained by kicking Scots pines on the outside of a clump, when the moth glides to the ground; comes to light.

1419(18). **D. mutatella** Fuchs
O. On *Pinus sylvestris*.
L.
P.
I. Differences between the life histories of this and the previous species are not yet known; Ford thought it probable that the cone-feeder and shoot-feeder were distinct. (16) *D. splendidella* (Herr.-Schäff.) does not occur in Britain.

1420(14). **Epischnia bankesiella** Rich.
O. On *Inula crithmoides* growing amongst limestone rocks.
L. 8-5. In a gallery extending from the base of the plant to the top of the shoots, which are often spun together conspicuously.
P. 6-7. In a cocoon at base of the foodplant.
I. 7. Flies at dusk and later comes to light.

1421(15). **Hypochalcia ahenella** ([D. & S.])
O. On *Helianthemum* or *Artemisia campestris*.
L. 7-9.

P.
I. 6-7. A downland species, easily disturbed in the daytime; flies at dusk and later comes to light.

1422(43). **Myelois cribrella** (Hübn.)
O. On *Cirsium vulgare, C. eriophorum, Onopordon* or, less often, *Arctium lappa*, preferring large plants.
L. 8-4. At first on the seeds; enters the stem in the autumn and feeds on the pith, hibernating when nearly full-grown; often several in one stem.
P. 5-6. In the stem, near its circular exit-hole.
I. 7-8. Comes to light; rests on thistle-heads in the evening.

1423(42). **M. cirrigerella** (Zinck.) *Ent.Rec.* **47**:124(1935)
O. On *Knautia* or *Centaurea scabiosa*.
L. 8-5. On the seeds, from a silken web, several larvae sometimes being found in one flower-head; hibernates, sometimes twice, in its cocoon.
P. 6-7. On or in the earth at the root of the foodplant.
I. 7-8. Frequents downland, resting in the daytime under or on the flower-heads of the foodplant; flies at night and comes to light.

1424(45;46). **Ectomyelois ceratoniae** (Zell.) (*phoenicis* Durr.)
O.
L. 5-7. On locust beans, dried fruit and nuts, and tamarind pods.
P. 6-7.
I. 8-9. Sometimes plentiful in warehouses where it has no doubt been imported with the foodstuff.

1425(34). **Assara terebrella** (Zinck.) *Ent.Rec.* **47**:51(1935)
O.
L. 9-6. In a cone of *Picea abies*, inhibiting its development and causing it to fall prematurely; possibly living two years.
P. 6-7. In the cone.
I. 6-8. Best obtained by gathering fallen cones 50-100 mm in length during the winter or spring. The moth flies at dusk and later comes to light.

1426(3). **Pempelia dilutella** (Hübn.)
O. On *Thymus drucei*, apparently always near a nest of the yellow ant *(Lasius flavus)*.
L. ?8-6. Probably from the autumn in a gallery of silk mixed with debris from the ants' nest, extending from an object on the ground to the leaves of the foodplant.
P. 7. In a cocoon in the gallery or amongst spun leaves.
I. 7-8. Frequents downland and is easily disturbed by day; flies towards dusk and later comes to light.

1427(4). **P. ornatella** ([D. & S.])
O. On *Thymus drucei*.
L. 5-6. In a tube-like web amongst the roots.
P.
I. 7. Flies at dusk and possibly later.

1428(2). **Gymnancyla canella** ([D. & S.]) *Ent.* **60**:279(1927)
O. On *Salsola kali* or *Atriplex* spp.
L. 8-10. When young, mining the stem or, more usually, a side-shoot, causing the tip to wither. In September it may leave the stem and feed on the unripe seeds from a web rendered conspicuous by sand-grains.
P. 10-7. In a cocoon in sand near the foodplant; sometimes over-wintering twice.
I. 7-8.

1429(12). **Alispa angustella** (Hübn.)
O. 6-7;9-10. On the berries of *Euonymus europaeus*.
L. 8;10. In the berries, when nearly full-grown spinning two or more together and exuding much frass.
P. 9;11-6. In rotten wood.
I. 6-7;9. Flies at night and comes to light. The moth emerges freely during the winter if the pupae are kept indoors. The earlier larval brood occurs only in favourable seasons.

1430(–). **Mussidia nigrivenella** (Rag.)
O.
L. On cereals and cacao beans.
P.
I. An African species taken in a London warehouse in 1950.

1431(31). **Ancylosis oblitella** (*Heterographis oblitella*) (Zell.) *Ent.Gaz.* **8**: 37(1957)
O. Hatches in 13 days.
L. On Chenopodiaceae, especially in salt-marshes.
P. In a thick silken cocoon.
I. 6-8. Readily disturbed in the daytime; flies at night and comes to light. Formerly a rare migrant, it has been established along the Thames estuary since 1956 and is extending its range.

1432(41). **Nyctegretis achatinella** (Hübn.)
O. On *Ononis* spp., also on *Trifolium* spp., *Artemisia* spp. or *Sarothamnus*.
L. 8-5. In a silken tube beneath the foodplant.
P. 6-7. In a cocoon attached to a dead stem in the sand.
I. 7. Frequents coastal sandhills; flies at dusk and later comes to light and may be found resting on flowers of *Echium vulgare*, etc.

1433(33). **Euzophera cinerosella** (Zell.)
O. On *Artemisia absinthium*, hatching in about 14 days.
L. 9-5. In the root, later burrowing up into the stem.
P. 5-7. In the stem.
I. 6-8. Flies at dusk and later comes to light.

1434(32). **E. pinguis** (Haw.)
O. On *Fraxinus*, preferring pollarded trees.
L. 9-7. On living inner bark, throwing black frass from the entrance-hole; possibly living two years.
P. 7. In the larval burrow.
I. 7-8. Emerges 1900-2100 hrs, when it may be found drying its

wings; infests only certain trees, leaving the others unattacked. Flies at dusk and later comes to light.

1435(–). **E. osseatella** (Treits.) *Ent.Gaz.* **14**:100(1963)
O.
L. On potato tubers imported from abroad.
P.
I. A specimen was bred in 1962 from a larva found in a consignment of Egyptian potatoes.

1436(–). **E. bigella** (Zell.) *Ent.Gaz.* **19**:158(1968)
O.
L. In peaches.
P.
I. One bred in Edinburgh in 1955 from an imported Italian peach.

1437(25). **Ephestia elutella** (Hübn.)
O. On cacao, chocolate, cereals, tobacco, dried fruit, dried vegetables, walnuts or other nuts, seeds, chicory, cayenne pepper, sugar or other stored products; occasionally on dried animal materials; in haystacks. Eggs hatch in 5-15 days.
L. 8-4. At first, internally, if the foodstuff permits; later in a frass-covered web.
P. 4-6. In a cocoon amongst the foodstuff or, in a warehouse, in a crevice, frequently in the ceiling.
I. 6-10. Generally rests in dark corners of a warehouse, but has also been taken out-of-doors at flowers and light, and is sometimes abundant in old haystacks.

1438(24). **E. parasitella unicolorella** Staud. (*woodiella* R. & T.)
O.
L. 9-5. Probably on dry vegetable refuse, including dry berries and dead stems of ivy; overwinters in its cocoon.
P. 5. In a silken cocoon.
I. 6-9. Has been beaten from ivy and alder and has been taken in salt-marshes and at light.

1439(29). **E. kuehniella** Zell.
O. Laid on or near flour, hatching in three to six days.
L. 8-5. In a silken tube on flour; occasionally on other stored vegetable products or on dead insects.
P. 5-10. In a tough silken cocoon attached to a solid object near its food.
I. 7-10. A pest in almost every flour-mill; adults generally remain inside the mill, resting on walls and windows by day and flying at night.

1440(27). **E. cautella** (Walk.)
O. 7-4. On dried figs, currants, nuts, beans, etc.
L. 4-10. In a dense greyish web of silk mixed with frass.
P. 5-10. In a slight cocoon amongst or close to the foodstuff.
I. 5-10. Remains in the warehouse, sitting on walls or windows by day and flying at dusk.

1441(26). **E. figulilella** Gregs.
O.
L. 9-6. On dried figs, currants, etc.
P.
I. 7-10. Possibly in two broods. In warehouses, but has also been taken out-of-doors at light.

1442(28). **E. calidella** Guen.
O.
L. 9-5. On dried fruits, nuts, etc.
P. 5-9.
I. 7-10. In warehouses.

1443(23). **Plodia interpunctella** (Hübn.)
O. 6-7. Laid in clusters on or near stored cereals, dried fruits, nuts or some other stored vegetable product; on dried insects. Hatching in 2-14 days.
L. 8-6. In a silken web amongst the foodstuff.
P. 6-9. In a slight web near or amongst the foodstuff.
I. 6-10. Sometimes abundant in warehouses, where it rests on walls by day and flies by night. In heated warehouses there may be up to three overlapping broods.

1444(40). **Homoeosoma nebulella** ([D. & S.])
O. 8-9. On *Cirsium vulgare, Senecio jacobaea, Tanacetum vulgare* or *Leucanthemum vulgare*.
L.8-5. In the flower-head on the flowers and seeds; it probably overwinters in its cocoon.
P. 6-7. In a cocoon in the soil or in leaf-litter.
I. 6-8. Flies at dusk and later comes to flowers of ragwort and thistle and to light.

1445(35). **H. sinuella** (Fabr.)
O. On *Plantago lanceolata*; possibly on other *Plantago* spp.
L. 8-4. In the rootstock, causing the central leaves to droop in the autumn and stunting growth in the spring; larvae become full-grown from late autumn to early spring.
P. 4-7. In the rootstock.
I. 6-7. Flies from early evening onwards, later comes to light.

1446(38). **H. nimbella** (Dup.)
O. 8. On *Senecio jacobaea, Tanacetum parthenium, Anthemis* spp. or other Compositae.
L. 9-4. In the flower-head, sometimes gregariously; overwinters in its cocoon.
P. 5. In a thick silken cocoon.
I. 6-8. Flies at dusk.

1447(36). **Phycitodes binaevella** (*Rotruda binaevella*) (Hübn.)
O. On *Cirsium vulgare* or *Carduus pycnocephalus*.
L. 8-9. On the seeds, in a large cavity in the base of the flower-head; overwinters in its cocoon.
P. 6-7. In a cocoon spun in leaf-litter.
I. 7. Rests in the daytime under the flower-head of the food-plant, but on top at night; flies at dusk and later visits flowers of thistle and ragwort.

1448(39). **P. saxicola** (Vaugh.)
O.
L. 7;9. Except that it appears to be bivoltine, the early stages have not been distinguished from those of *Homoeosoma nimbella* (no. 1446).
P. 5;7.
I. 6;8. Flies freely at dusk.

1449(37). **P. carlinella** (Hein.) (*cretacella* (Rössl.))
O. On *Senecio jacobaea, Tanacetum vulgare* or *Achillea millefolium*.
L. 6;8-4. In the upper stem and leaf-axils, later on the flowers in a web; sometimes gregarious; autumnal larvae overwinter in their cocoons.
P. 4-5;6-7. In a cocoon, generally away from the foodplant.
I. 5;7. Flies at night, visiting ragwort blossom and light. Possibly there is a prolonged emergence period rather than two broods.

1450(44). **Apomyelois bistriatella neophanes** (Durr.)
O. On *Daldinia concentrica* growing on dead birch, less often on gorse or other plants, especially on burnt stems.
L. 8-5. Within the fungus; if, when the fungus is broken off, white silk threads are seen, there should be a larva in the fungus or the wood nearby.
P. 5-6. In the stem or under the bark of the host tree, near the fungus.
I. 6-7. Flies in sunshine; in dull weather may be obtained by tapping old gorse stems, when it drops to the ground.

PTEROPHORIDAE

1451(212). **Agdistis staticis** Mill.
O. On *Limonium binervosum*, usually on the upper surface of a leaf, but sometimes on the under surface or on a flower.
L. 8;10-5. On the leaves, preferring sheltered situations.
P. 6-7;9. On a leaf, adjacent herbage or bare rock.
I. 7-8;10. Sometimes rests on the bare cliff-face.

1452(211). **A. bennetii** (Curt.)
O. On a small central leaf of *Limonium vulgare*, more usually on solitary plants.
L. 7;9-5. On the leaves, resting on the underside; large arcs eaten from the leaves indicate the presence of a larva.
P. 6;7-8. On the upper surface of a leaf.
I. 6;8-9. The imagines do not fly till 2000-2100 hrs, but may sometimes be disturbed on warm days; they come to light and sugar.

1453(180). **Oxyptilus pilosellae** (Zell.)

O. 8. On the underside of a leaf of *Hieracium pilosella*, hatching in about 10 days.

L. 8-5. Probably mining leaves before hibernation; in the spring on young leaves in the heart of the plant.

P. 6. In a cocoon under a leaf.

I. 7-8. Frequents downland; rests on grass-stems in the afternoon, and flies freely in the evening.

1454(181). **O. parvidactylus** (Haw.)

O. 7. Laid singly on the upper surface of a leaf of *Hieracium pilosella*.

L. 8-5. In spring in the central rosette, causing it to wither slightly.

P. 5-6.

I. 6-7. Flies actively in afternoon sunshine and at dusk; visits flowers of wild thyme.

1455(179). **Crombrugghia distans** (Zell.)

O. On *Hieracium pilosella*; probably also on *Crepis capillaris* or *Picris hieracioides*.

L. 7;9-?4. On the flowers; the second brood may hibernate in the stems.

P.

I. 5-6;7-8. Frequents sandy coasts, the Breck sands and, less often, chalk downland; is readily disturbed in hot afternoon sunshine and flies freely at dusk.

1456 (–). **C. laetus** (Zell.) *Ent.Rec.* **75**:11(1963); *Proc.Brit.ent.nat.Hist. Soc.* **2**(2):39(1969)

O.

L.

P.

I. Twice recorded in Britain; possibly an immigrant.

1457(178). **Buckleria paludum** (Zell.)

O. On the underside of a leaf, a petiole or dead flower-stalk of *Drosera rotundifolia*.

L. 7;9-5. On the leaves, leaving piles of green frass.

P. 6;7. Under a leaf or on adjacent herbage.

I. 6;8. Flies in the evening and after dark; can be disturbed in the daytime.

1458(182). **Capperia britanniodactyla** (Gregs.)

O. On *Teucrium scorodonia*.

L. 9-5. Overwinters under a leaf; in the spring the larva nibbles halfway through the stem causing the upper part of the plant to wither, and feeds on the tip of the withered part.

P. 6. On the stem under the withered part, under a leaf or on adjacent herbage.

I. 6-7. Flies at dusk and later comes to light; not easily disturbed by day.

1459(205). **Marasmarcha lunaedactyla** (Haw.)

O. On the underside of a leaf of *Ononis repens* or *O. spinosa*.

L. 5-6. On the upperside of the leaves.

P. 6-7. On a leaf or the stem of the foodplant.

I. 7. Easily disturbed in the daytime; flies at dusk and after dark.

1460(183). **Cnaemidophorus rhododactyla** ([D. & S.])

O. 8. On the underside of a leaf of *Rosa canina*, hatching in about 14 days.

L. 8-6. In the autumn bores into the soft part of a stem, where it perhaps overwinters; in the spring on flower-buds and shoots, spinning a leaf to a bud.

P. 6-7. On the foodplant close to where the larva has fed, under a few strands of silk.

I. 7-8. Flies at night and comes to light.

1461(185). **Amblyptilia acanthadactyla** (Hübn.)

O. On *Stachys sylvatica, Ononis* spp., *Mentha* spp., *Geranium* spp., *Chenopodium* spp., *Calluna* or *Erica*.

L. 6;8. On the flowers and unripe seeds.

P. 6-7;8-9. On the flower-stalk, amongst the flowers or seeds, resembling a dead flower.

I. 7;9-5. Flies at dusk and at night; the autumn generation comes to ivy-bloom.

1462(184). **A. punctidactyla** (Haw.) (*cosmodactyla* (Hübn.))

O. On *Stachys sylvatica, Aquilegia, Erodium cicutarium, Primula* spp. or *Geranium* spp.

L. 6;8. In the seed-vessels.

P. 6-7;8-9. On the foodplant.

I. 7;9-5. Sometimes flies in sunshine but normally at night, when it visits ragwort-blossom.

1463(186). **Platyptilia tesseradactyla** (Linn.)

O. On *Antennaria*.

L. 8-5. In the flower-stem and shoots, aborting the growth.

P. 5.

I. 5-6. Easily disturbed; flies in evening sunshine, and rests on flower-heads of the foodplant.

1464(187). **P. calodactyla** ([D. & S.])

O. 7-8. On a leaf of *Solidago virgaurea*.

L. 8-5. In the stem, in spring sometimes causing slight withering of the leaves.

P. 5. In the larval burrow.

I. 6-7. Easily disturbed in the daytime, when it flies swiftly; flies after dark and comes to light.

1465(188). **P. gonodactyla** ([D. & S.])

O. 6-7;9. Laid singly on the underside of a leaf of *Tussilago*; possibly also on *Petasites* spp.

L. 7-8;9-5. Summer brood first mines leaves, then feeds externally on their undersides. Autumn larvae at first mine leaves, then enter the stems in which they overwinter; in spring they eat the buds, flowers and seeds, spinning the seed-heads so that the down is retained and the head droops.

P. 5-6;8-9. In spring in the seed-head; in summer under a leaf.

I. 6;9. Flies at dusk and later comes to light.

1466(189). **P. isodactylus** (Zell.)

O. 6-7;9. On the underside of a leaf of *Senecio aquaticus*.

L. 7-8;9-5. In the stem, exuding frass at the axil of a leaf; also in the flower-head and on the leaves. Autumn larvae overwinter in the stems.
P. 5-6;8. In the stem or under a web on the upperside of a leaf of the foodplant or adjacent herbage.
I. 6;8-9. Flies at dusk and later sometimes comes to light.

1467(190). **P. ochrodactyla** ([D. & S.])
O. 8-9. On a floret of *Tanacetum vulgare*.
L. 9-6. Mines down a stem to overwinter in the root; in the spring burrows up a new shoot, exuding frass at the joints and causing the upper part to droop.
P. 6-7. On the stem of the foodplant.
I. End 7-8. Rests on the flowers after dark.

1468(191). **P. pallidactyla** (Haw.)
O. 7-8. On a flower-head of *Achillea millefolium* or *A. ptarmica*; possibly sometimes on *Tanacetum vulgare*.
L. 9-6. Mines down a stem to overwinter in the root; in spring, mines a succession of young shoots, causing drooping.
P. 6. On the stem or a leaf of the foodplant.
I. 6-7. Flies at dusk and later comes to light; easily disturbed in the daytime.

1469(206). **Stenoptilia graphodactyla** (Treits.) (*pneumonanthes* (Buttner))
O. 7-8;9. On a flower, bud or stem of *Gentiana pneumonanthe*.
L. 8;10-5. On the developing seeds; autumnal larvae overwinter in the stems and feed on the central shoots in the spring.
P. 6-7;8-9. On the stem or a leaf of the foodplant.
I. 7;9. Frequents boggy heaths; flies in the evening; may be disturbed by day.

1470(210). **S. saxifragae** Fletch.
O. Laid singly on the underside of a leaf of *Saxifraga hypnoides* or possibly, other *Saxifraga* spp., hatching in about 10 days.
L. 7-5. Mining leaves before and after hibernation, which takes place in the stem; when nearly full-fed it feeds externally on the shoots.
P. 6. On the foodplant.
I. 7-8. Flies at dusk and later comes to light. Some larvae feed up rapidly and produce adults in September.

1471(207). **S. zophodactylus** (Dup.)
O. 7;9. On a leaf (July) or seed-vessel (September) of *Centaurium erythraea, Blackstonia* or *Gentiana* spp.
L. 6;8. First mining a basal leaf, then successive buds, causing them to turn yellowish, and finally externally on the flowers and seeds.
P. 7;9. On the stem of the foodplant.
I. 7;9. In two overlapping broods. Easily disturbed in the daytime; flies at dusk.

1472(208). **S. bipunctidactyla** (Scop.)
O. 6;9. On *Succisa, Knautia*, or *Scabiosa*.
L. 6-7;9-4. In summer and autumn on the flowers; after hibernation, in the young shoots and leaf-buds, filling them with frass.

P. 5;7-8. On the foodplant or adjacent vegetation.
I. 5-6;8-10. In two or three overlapping broods. Easily disturbed in the daytime; flies at dusk and after dark.

1473(209). **S. pterodactyla** (Linn.)
O. 8. On the underside of a leaf of *Veronica chamaedrys*.
L. 8-5. At first in the stem of a terminal shoot, killing the tip, then hibernating in the stem; in the spring, exposed on the flowers.
P. 5-7. On the foodplant or adjacent herbage.
I. End 6-early 8. Easily disturbed in the daytime; flies at dusk and later comes to light and visits flowers.

1474(192). **Pterophorus tridactyla** (Linn.) (*tetradactyla* (Linn.))
O. On *Thymus drucei*.
L. ?9-6. On the leaves.
P. 6-7. On the foodplant.
I. 6-8. Easily disturbed in the daytime; flies at dusk and after dark.

1475(196A). **P. fuscolimbatus** (Dup.) (*icterodactylus phillipsi* Huggins) *Ent. Gaz.* **4**:139-140(1953); **6**:124-126(1955)
O.
L. Probably on Labiatae; *on *Lavandula* spp.
P.
I. 5-7. The life history is not known in Britain.

1476(193). **P. baliodactylus** Zell.
O. 8. On *Origanum*.
L. 8-6. Hibernates when small; in spring bites half through a stem near the top and feeds on the leaves thus caused to wither.
P. 6-7. On the stem or a leaf of the foodplant.
I. 7-8. Easily disturbed in the daytime; flies at dusk and after dark.

1477(194). **P. pentadactyla** (Linn.)
O. 7-8. On *Convolvulus* or *Calystegia* spp.
L. 8-5. Hibernates when small; in spring eats the leaves and flowers.
P. 5-6. Under a leaf of the foodplant.
I. 6-7. Easily disturbed in the daytime; flies at dusk and later comes to light. Sometimes some of the August larvae feed up quickly and produce a second generation of adults in September.

1478(195). **P. galactodactyla** ([D. & S.])
O. 8. On *Arctium lappa*.
L. 8-5. Hibernates when small; in spring, eats out conspicuous, oval-shaped holes in the leaves, moving to younger leaves as the plant grows.
P. 6. On the underside of a leaf, generally close to a rib near the the base.
I. 6-7. Flies at dusk and at night; sometimes rests by day on the upperside of burdock leaves.

1479(196). **P. spilodactylus** Curt.
O. 8. On *Marrubium*.
L. 8-6. Hibernates when small; in spring feeds on the leaves, mainly from the upperside.

P. 6-7. On the upperside of a leaf.
I. 7-8. Some larvae feed up quickly and produce a small second generation in September; larvae, pupae and imagines are often found at the same time on one plant.

1480(197). **Pselnophorus heterodactyla** (Müll.) (*brachydactyla* (Koll.))
O. On the underside of a leaf of *Mycelis*, hatching in about a month.
L. 8-5. When young, on the lower cuticle of a leaf, making conspicuous blotches; hibernates in a dead leaf; in spring, bites through the midrib of a leaf so that the tip hangs down.
P. 6. On or near the foodplant.
I. 6-7. Frequents the darker parts of large woods; fond or resting in the evening on the flowers of yellow bird's-nest (*Monotropa hypopitys*).

1481(198). **Adaina microdactyla** (Hübn.)
O. On *Eupatorium*.
L. 7;9-10. In the stem, usually in the axils of a bunch of flower-stalks, making a gall from which frass exudes and causing the flowers to droop.
P. 7-8;11-5. In the stem, usually below the gall; some second-generation larvae do not pupate until the spring.
I. 5-6;8. Readily disturbed in the daytime; flies at dusk and after dark. Stems of the foodplant having small holes, if gathered in the winter, will usually produce many imagines.

1482(199). **Leioptilus lienigianus** (Zell.)
O. On *Artemisia vulgaris*; also on *Leucanthemum vulgare*, *Solanum* spp., etc.
L. 5-6. In a spun pocket on a leaf, blotching it; making several such pockets.
P. 6-7. On a leaf, the lower part of the stem or adajcent herbage.
I. 7. Flies at dusk and after dark.

1483(200). **L. carphodactyla** (Hübn.)
O. 6-7;9. On *Inula conyza* or *Carlina*.
L. 7-8;9-5. In the summer, in the flowers causing some of the petals to protrude; autumn larvae hibernate when small and in the spring feed in the heart of the plant, aborting its growth.
P. 6;8. On the flowers or leaves of the foodplant.
I. 6;8-9. Flies at dusk; not easily disturbed by day.

1484(201). **L. osteodactylus** (Zell.)
O. 8. On *Solidago virgaurea* or *Senecio* spp.
L. 9-4. On the flowers and seed-head; hibernates full-fed amongst leaf-litter.
P. 5-6. On or near the foodplant.
I. 7. Easily disturbed in the daytime; flies at dusk and later comes to light.

1485(–). **L. chrysocomae** Rag. (*bowesi* Whalley) *Ent. Gaz.* **11**:29-30(1960); *Ent.* **95**:9-10(1962)
O. 8. On *Solidago virgaurea, Senecio* spp. or *Aster* spp.
L. 9-4. On the flowers and seed-head; hibernates full-fed amongst leaf-litter.

P. 5-6. On or near the foodplant.
I. 7. Easily disturbed in the daytime; flies at dusk and later comes to light. The early stages have not been satisfactorily distinguished from those of the preceding species.

1486(202). **L. tephradactyla** (Hübn.)
O. On *Solidago virgaurea* or *Aster* spp.
L. 9-5. On the leaves, eating out holes at night; when not feeding, rests on the underside of the leaf. Hibernates from August about one-third grown.
P. 5-6. On the stem or a leaf of the foodplant.
I. 6. Easily disturbed in the daytime; flies at dusk and later comes to light.

1487(203). **Oidaematophorus lithodactyla** (Treits.)
O. On *Pulicaria dysenterica* or *Inula conyza*.
L. 5-6. At first in the young shoots; later, exposed on the leaves which it riddles with holes.
P. 6-7. On the upperside of a leaf of the foodplant.
I. 7-8. Flies at dusk and after dark.

1488(204). **Emmelina monodactyla** (Linn.)
O. 5. On *Convolvulus* or *Calystegia* spp., especially *C. sepium*; occasionally on other herbaceous plants.
L. 6-8. On the young shoots.
P. 8-9. Usually away from the foodplant, often on a fence.
I. 9-4. Flies at dusk and later comes to light; visits ivy-bloom, ripe blackberries, etc. in the autumn and sallow-bloom in the spring.

INDEX TO FOODPLANTS AND OTHER FOOD SUBSTANCES AND SITUATIONS

Both Latin and English names are those given in "English Names of Wild Flowers: a List recommended by the Botanical Society of the British Isles", by J.G. Dony, C.M. Rob and F.H. Perring, 1974.

INDEX OF FAMILIES, GENERA AND SPECIES

Species are indicated by their serial numbers, a family or genus by the serial number of the first (or only) species in that family or genus. Synonyms are in italics. Names not allocated serial numbers are listed under the number of the preceding species placed in square brackets.